International Perspectives on Mathematics Curriculum

edited by

Denisse R. Thompson
University of South Florida

Mary Ann Huntley
Cornell University

Christine Suurtamm
University of Ottawa

INFORMATION AGE PUBLISHING, INC.
Charlotte, NC • www.infoagepub.com

Library of Congress Cataloging-in-Publication Data

A CIP record for this book is available from the Library of Congress
http://www.loc.gov

ISBN: 978-1-64113-043-1 (Paperback)
    978-1-64113-044-8 (Hardcover)
    978-1-64113-045-5 (ebook)

Copyright © 2018 Information Age Publishing Inc.

All rights reserved. No part of this publication may be reproduced, stored in a retrieval system, or transmitted, in any form or by any means, electronic, mechanical, photocopying, microfilming, recording or otherwise, without written permission from the publisher.

Printed in the United States of America

International Perspectives on Mathematics Curriculum

A volume in
Research in Mathematics Education
Denisse R. Thompson, Mary Ann Huntley,
and Christine Suurtamm, *Series Editors*

CONTENTS

Preface vii

1 What Might Be Learned From Examining Curricular
   Perspectives Across Countries? 1
   Christine Suurtamm, Mary Ann Huntley, and Denisse R. Thompson

2 Primary School Mathematics in the Netherlands: The
   Perspective of the Curriculum Documents 9
   Marc van Zanten and Marja van den Heuvel-Panhuizen

3 Curriculum in France: A National Frame in Transition 41
   *Ghislaine Gueudet, Lætitia Bueno-Ravel, Simon Modeste,
   and Luc Trouche*

4 Mathematics Curriculum: The Case of Finland 71
   Kirsti Hemmi, Heidi Krzywacki, and Anna-Maija Partanen

5 Curriculum in Canada: A Fractal Interpretation Using the
   Case of Alberta 103
   Elaine Simmt

6 Mathematics Curriculum in the United States: New Challenges
   and Opportunities 133
   Janine Remillard and Luke Reinke

7   A South African Perspective on the Mathematics Curriculum: Towards A Transcending Metacognitive Ideology 165
    Divan Jagals and Marthie van der Walt

8   Discussing the Mathematics Curriculum in Brazil 191
    Celi Espasandin Lopes and Regina Célia Grando

9   The Korean Mathematics Curriculum: Characteristics and Challenges 211
    Kyeong-Hwa Lee, JinHyeong Park, and Na-Young Ku

10 Transcending Boundaries: What Have We Learned? 229
    Mary Ann Huntley, Christine Suurtamm, and Denisse R. Thompson

    About the Editors 241

    About the Contributors 243

PREFACE

Since the 1960s, there have been numerous international comparative studies undertaken by the International Association for the Evaluation of Educational Achievement (IEA). Such studies have often involved large-scale testing on content common to participating countries but have also involved the analysis of curriculum, specifically in the Second International Mathematics Study (SIMS; Travers & Westbury, 1989) and the Third International Mathematics and Science Study (TIMSS; Schmidt, McKnight, Valverde, Houang, & Wiley, 1997). These studies have led to conceptualizations of curriculum at different levels, including the *intended curriculum*, the *implemented curriculum*, and the *attained curriculum* (Travers, 1992). Others have since expanded on these levels and considered factors that influence curriculum at each level (Remillard & Heck, 2014). But despite the importance of and interest in curriculum, research on mathematics curriculum is a relatively recent phenomenon (Li & Lappan, 2014).

School mathematics is part of the educational experience around the world. As noted by Schmidt et al., some of students' experiences will be common around the world, and others will be quite different.

> *Curriculum* is the most fundamental structure for these experiences. It is a kind of underlying "skeleton" that gives characteristic shape and direction to mathematics instruction in educational systems around the world.... The plan that expresses these aims and intentions, which takes them from vision to implementation, and serves as the broad course that runs throughout for-

mal schooling, is *curriculum*. Curriculum provides a basic outline of planned and sequenced educational opportunities. (1997, p. 4)

To understand learning differences, it is important to understand different opportunities to learn that may be linked to differences in curriculum opportunities or fundamental educational structures. That is, a careful look at potential differences in the intended curriculum across countries can shed insight into priorities of a country in terms of mathematical learning, and how cultural contexts influence those priorities. This book provides one such opportunity to consider how diverse countries from five continents (Africa, Asia, Europe, North America, South America) position the mathematics curriculum within their broad educational systems.

OUR CALL FOR MANUSCRIPTS

When we first conceptualized this volume, we were responding to the call by Kulm and Li (2009) that it is important to study curriculum "to reveal the expectations, processes and outcomes of students' school learning experiences that are situated in different cultural and system contexts.... Further studies of curriculum practices and changes are much needed to help ensure the success of educational reforms in the different cultural and system contexts" (p. 709). We were particularly interested in a lens that focused on different ways that curriculum might be developed or understood, or perhaps implemented, in various educational jurisdictions or countries.

We invited curriculum scholars from a representative sample of countries to contribute to the volume. Although we gave them much latitude in developing their chapters, we did provide the following guiding questions, which were neither meant to be exhaustive nor limiting:

- How is curriculum defined in your country?
- To what extent does the curriculum reflect current research and thinking in mathematics education?
- What is the underlying framework of the curriculum?
- How is the curriculum organized?
- What vision of mathematics education is portrayed? What does the curriculum value?
- How is the curriculum vision shared among different stakeholders (e.g., teachers, parents, the general public)?
- What are some of the challenges faced in the implementation of the curriculum?

We hope readers will agree that the chapter authors have given us much to think about relative to these issues.

POTENTIAL AUDIENCE FOR THE BOOK

The volume has the potential to be applicable to curriculum researchers, both in mathematics and even more generally, including those interested in comparative issues in curriculum and education. We hope the volume will stimulate conversation about curriculum among the wider mathematics education community and provide a means to consider similarities and differences in curriculum, and how those similarities and differences may relate to cultural and societal differences across countries. The volume might be used by policy makers interested in comparing mathematics curriculum around the world as well as by professors as a text in a graduate course on curricular issues.

We believe this volume is a natural complement to several other volumes published by Information Age that focus on issues in mathematics curriculum, including the following:

- *The Intended Mathematics Curriculum as Represented in State-Level Curriculum Standards: Consensus or Confusion?* (Reys, 2006), which explored the placement of mathematics topics in the intended curriculum in grades K–8 in the United States;
- *Mathematics Curriculum in Pacific Rim Countries—China, Japan, Korea, and Singapore: Proceedings of a Conference* (Usiskin & Willmore, 2008), investigating the mathematical frameworks used to design curriculum in each of the named countries;
- *Future Curricular Trends in School Algebra and Geometry: Proceedings of a Conference* (Usiskin & Andersen, 2010), exploring curricular issues in two specific content areas that are included in most school experiences around the world;
- *Approaches to Studying the Enacted Mathematics Curriculum* (Heck, Chval, Weiss, & Ziebarth, 2012) that identifies instruments used in various curriculum endeavors to study the enacted curriculum;
- *Enacted Mathematics Curriculum: A Conceptual Framework and Research Needs* (Thompson & Usiskin, 2014) that focuses on research related to the enacted curriculum and expands on discussions at a conference related to researching the enacted curriculum; and
- *Digital Curricula in School Mathematics* (Bates & Usiskin, 2016) that considers how researchers in different countries are addressing challenges surrounding digital curriculum in the current educational and globalized environment.

STRUCTURE OF THE VOLUME

The volume consists of ten chapters. Chapter 1 provides an orientation to the volume, and gives the reader possible questions to guide reflection while reading the various chapters.

Chapters 2–9 are the heart of the volume. Each chapter outlines the mathematics curriculum in one specific country (the Netherlands, France, Finland, Canada, United States, South Africa, Brazil, or South Korea, respectively), telling the curricular story from the perspective of the authors, and raising issues and challenges each country faces.

Chapter 10 provides a look back at the volume, with some of our views on the guiding questions from Chapter 1, including some facets that we found particularly interesting or unusual. We end the chapter with possible ideas for future research related to the messages within the various chapters. The volume concludes with brief biographies of the editors and authors.

ACKNOWLEDGEMENTS

We extend our thanks to the many authors for their hard work in writing chapters for the volume. We appreciate all their efforts in generating drafts and responding to our questions and edits. We are also grateful that the authors who attended ICME-13 (July 2016 in Hamburg, Germany) took time to meet with us to discuss their chapters.

In addition, we extend our thanks to George Johnson, President of Information Age Publishing, for his willingness to publish the volume. As co-editors of the series, *Research in Mathematics Education*, we appreciate the freedom provided to cultivate this volume related to our special interest in mathematics curriculum.

REFERENCES

Bates, M., & Usiskin, Z. (Eds.). (2016). *Digital curricula in school mathematics.* Charlotte, NC: Information Age.

Heck, D., Chval, K., Weiss, I., & Ziebarth, S. W. (Eds.). (2012). *Approaches to studying the enacted mathematics curriculum.* Charlotte, NC: Information Age.

Kulm, G., & Li, Y. (2009). Curriculum research to improve teaching and learning: national and cross-national studies. *ZDM: The International Journal on Mathematics Education, 41,* 709–715.

Li, Y., & Lappan, G. (Eds.). (2014). *Mathematics curriculum in school education.* Dordrecht, the Netherlands: Springer.

Remillard, J. T., & Heck, D. J. (2014). Conceptualizing the enacted curriculum in mathematics education. In D. R. Thompson & Z. Usiskin (Eds.), *Enacted mathematics curriculum: A conceptual framework and research needs* (pp. 121–148). Charlotte, NC: Information Age.

Reys, B. (Ed.). (2006). *The intended mathematics curriculum as represented in state-level curriculum standards: Consensus or confusion?* Charlotte, NC: Information Age.

Schmidt, W. H., McKnight, C. C., Valverde, G. A., Houang, R. T., & Wiley, D. E. (1997). *Many visions, many aims: A cross-national investigation of curricular intentions in school mathematics.* (Volume 1). Dordrecht, the Netherlands: Kluwer.

Thompson, D. R., & Usiskin, Z. (Eds.). (2014). *Enacted mathematics curriculum: A conceptual framework and research needs.* Charlotte, NC: Information Age.

Travers, K. J. (1992). Overview of the longitudinal version of the Second International Mathematics Study. In L. Burstein (Ed.), *The IEA study of mathematics III: Student growth and classroom processes* (pp. 1–14). Oxford, England: Pergamon Press.

Travers, K. J., & Westbury, I. (Eds.). (1989). *The IEA study of mathematics I: Analysis of mathematics curricula.* Oxford, England: Pergamon Press.

Usiskin, Z., Andersen, K., & Zotto, N. (Eds.). (2010). *Future curricular trends in school algebra and geometry: Proceedings of a conference.* Charlotte, NC: Information Age.

Usiskin, Z., & Willmore, E. (Eds.). (2008). *Mathematics curriculum in Pacific Rim countries—China, Japan, Korea, and Singapore: Proceedings of a conference.* Charlotte, NC: Information Age.

CHAPTER 1

WHAT MIGHT BE LEARNED FROM EXAMINING CURRICULAR PERSPECTIVES ACROSS COUNTRIES?

Christine Suurtamm
University of Ottawa

Mary Ann Huntley
Cornell University

Denisse R. Thompson
University of South Florida

This volume on international perspectives on mathematics curriculum responds to calls for cross-national curriculum studies to better understand curriculum practices and changes, and to "help ensure the success of educational reforms in the different cultural and system contexts" (Kulm & Li, 2009, p. 709). The book includes contributions from authors in Brazil, Canada, Finland, France, the Netherlands, South Africa, South Korea, and the United States to better understand curriculum practices in those countries, and hence, to be able to understand the meaning, development, frameworks, content, and implementation of mathematics curriculum from an international perspective. It is understood that these eight countries are a small sample of the world arena and thus, this is not a comprehensive study of international mathematics curriculum. Nevertheless, these authors present a variety of perspectives that help us think more deeply first and foremost about mathematics curriculum, and secondarily, about mathematics teaching and learning. Although we, as editors, provided some guidelines and guiding questions to the authors of the chapters, each chapter is unique in the way its curriculum story is told. Readers should keep in mind that each chapter offers one perspective on curriculum in that country. If a different set of authors had written the chapter, a different, yet presumably complementary story would be told about curriculum development in that country.

Curriculum can be defined in a variety of ways. Curriculum might be viewed as a body of knowledge, a product, a process, or praxis (Walker, 1982). Curricula can differ as they are conceptualized from a variety of theoretical perspectives and address the needs of teachers, students, and the context of schooling (Eisner, 1985). The study of curriculum from an international perspective discloses the variety of expectations, processes, and intended student outcomes that are situated in different educational systems and cultural contexts (Kulm & Li, 2009). Furthermore, the history and evolution of mathematics curriculum in varying contexts reflects the efforts to enhance mathematics teaching and learning. Examining this evolution with an international lens reveals different and evolving perspectives of *what is mathematics, who learns mathematics, what mathematics is important to learn, ways that students learn,* and *ways to teach to enhance that learning* (Li & Lappan, 2014). Furthermore, the chapters in this volume recognize that mathematics curricula evolve within a context that includes political, linguistic, cultural, and ideological influences.

We define *curriculum* more broadly than a written document that represents the set of standards or outcomes that define what mathematics is to be taught. Although that set of standards is included as part of what we consider as curriculum, we also consider curriculum to be the messages inherent in mathematics curriculum documents and resources, how these standards are understood by a variety of stakeholders, and ultimately how they are enacted in classrooms. The chapters in this book consider the variety of ways that curricula are developed, understood, and implemented in different jurisdictions and countries.

In this book, readers see the multiple facets of curriculum—the written curriculum, the intended curriculum, the enacted curriculum, and, in some cases, the experienced curriculum (Stein, Remillard, & Smith, 2007). Generally, the *written curriculum* refers to curriculum as it exists on the printed page, which might include the set of standards or outcomes as described in curriculum policy documents. The *intended curriculum* focuses on what is hoped to be attained through implementing the curriculum. Hence, the intended curriculum includes the messages that are within the curriculum that could be either explicit or implicit or both, such as underlying theoretical frameworks, or suggested teaching practices. The term *intended curriculum* might also include teacher-intended curriculum, meaning a focus on the teacher's interpretation of the curriculum and the teacher's decisions and intentions in implementing the written curriculum, as described by Remillard and Heck (2014a, 2014b). The *enacted curriculum* is viewed as the learning experiences jointly created by students and teachers and includes teachers' decisions in operationalizing the written and intended curricula (Cal & Thompson, 2014).

QUESTIONS TO CONSIDER

The chapters in the volume present different perspectives on the written, intended, and enacted curriculum as well as a discussion of the contextual factors that influence these components of curriculum. In reading and reviewing the chapters, we encourage the reader to consider the following questions, recognizing that the questions and responses overlap, are intertwined, and influence one another.

How is Curriculum Defined in Each Country? What Does it Look Like?

Curriculum takes on a variety of meanings in different contexts. In some cases, it is seen as a set of standards that guide the teaching of mathematics; in other cases, it might be a series of resources. The curriculum may also be descriptive, prescriptive, or somewhere in between.

How is the curriculum organized? There may be general standards that cut across all content areas or standards written specifically for each content area. Consider whether the curriculum is organized by mathematical domains, grade levels, and/or different streams for different students.

What content is presented? Does the curriculum contain process standards as well as content standards, noting that process standards are called different names in different countries, such as mathematical processes, practices, or competencies?

What Does the Curriculum Development Process Look Like?

Consider who takes part in the development of curriculum. Many different players might help to guide or direct the development of mathematics curriculum documents. In some cases, the curriculum represents a political agenda and hence, is guided by policy makers. In other cases, curriculum development might involve a variety of stakeholders that include researchers, parents, teachers, community representatives, and students. Consider how the curriculum is reviewed and revised. In some countries, there is a regular cycle of mathematics curriculum review and revision; in others, the revision of a curriculum may occur due to political shifts or reactions to results on international assessments.

What Role Does Research Play in Curriculum Development?

Mathematics education research can help to inform the development of mathematics curriculum. Emerging research in such areas as learning progressions, the use of technology, formative assessment to inform instruction, task design, and research in curriculum development itself are just a few of the many areas of mathematics education research that should be informing curriculum development. But to what extent is this research consulted?

What is the Underlying Framework of the Curriculum?

Although a theoretical framework is not always explicit in a curriculum document, quite often there is an implicit viewpoint of what mathematics is, how students learn mathematics, and what the best ways are to teach mathematics. One might want to consider Ernest's framework of the aims for teaching mathematics, which positions contrasting purposes to mathematics education such as learning basic skills for job preparation versus empowering the learner as a highly numerate and critical citizen (Ernest, 1991, 2014). Another way of looking at the assumptions and values of the curriculum is to step back and consider Eisner's five basic orientations to curriculum: development of cognitive processes, academic rationalism, personal relevance, social adaptation and social reconstruction, and curriculum as technology (Eisner, 1985). Another consideration is to examine the perspective on equity and inclusivity that might underlie the curriculum. In other words, what vision of mathematics and mathematics education is portrayed and what does the curriculum seem to value?

How Do Different Stakeholders View the Curriculum?

Consider the ways in which teachers view the curriculum. Do teachers see the curriculum as a framework to provide guidance, or as a strict document that needs to be followed in a step-by-step manner? How does the written curriculum become the teacher-intended curriculum? Consider the ways in which the teacher interprets the curriculum in making decisions in its implementation. In addition to teachers, consider also how parents and the general public view the curriculum. Is it accessible? Does it represent the values and knowledge deemed important by the community?

How Are the Political, Cultural, Linguistic, Social, and Ideological Characteristics of the Country Reflected in the Curriculum?

Mathematics curricula do not exist in a vacuum but are set within a context. How does the curriculum reflect the views and context of the country within which it is situated? What aspects of the curriculum reflect the cultural and social norms of the country? How does curriculum change as political views shift? In some cases, changing governments leads to educational and curricular change. What role does language play in the curriculum and its implementation, particularly in multi-lingual countries?

What is the Role of Assessment on the Written, Intended, and Enacted Curriculum?

Many countries report the influence of international assessments on curricular change, at all levels—written, intended, and enacted (Suurtamm et al., 2016). In some cases, scores on international assessments may cause significant changes to be made to the curriculum. As you examine the chapters, consider the ways in which large-scale assessments interact with curriculum. Consider whether large-scale assessments inhibit or drive curriculum reform. Do assessments play a role in defining what is important to be taught? How do assessments influence teachers' decisions about the intended and enacted curriculum?

The implementation of the curriculum may also include a monitoring component. If so, what does this look like? If an assessment is used to monitor curriculum implementation or how well students are learning the curriculum, how well does the assessment align with the curriculum?

What Does the Implementation of the Curriculum Look Like?

Evidence suggests there can be many differences between the written curriculum and the enacted curriculum (e.g., Grouws, Smith, & Sztajn, 2004;

Kilpatrick, 2003). Many questions arise such as: How are new curriculum ideas reflected in classroom practice? In what ways is the implementation of a curriculum supported? What does teacher professional learning (both preservice and inservice) look like concerning the curriculum and the implementation of new ideas? What are the curriculum developers' views on teachers' professional judgment as they transform curriculum?

What is the Role of Textbooks and Other Resources in Enacting the Curriculum?

Textbooks and teaching resources are often influencing factors between the written and enacted curriculum in mathematics education (Tarr, Chávez, Reys, & Reys, 2006). What role do resources such as textbooks play in transforming or influencing curriculum into practice? Consider whether the curriculum or textbooks drive what is taught in the classroom. How well do the textbooks and resources align with the curriculum? What opportunities do teachers have to use their own professional judgment to determine the resources they will use to implement the curriculum in their classroom?

What is the Mathematical Focus of the Curriculum?

This book is about *mathematics* curriculum. As you read the chapters, consider the mathematical content that is emphasized at various grade levels across the countries. As changes are made to the curriculum, are they incremental or large-scale changes? What is the role of mathematicians and other content experts in changes to the content of the curriculum? What is the balance between students learning concepts, skills, and procedures? To what extent are mathematical processes included in content standards, and are they separate from or intertwined with the content standards? What mathematical ideas are deemed as important in this curriculum?

SUMMARY

This book represents researched and thoughtful examinations of curriculum in eight different countries by mathematics education researchers who are situated in those countries. Each chapter provides us with insights into what mathematics is valued and taught, as well as how the curriculum evolved, who has been responsible for the development of curriculum, and ways that the curriculum is supported and enacted in the context of that country.

The purpose of the volume is not to do a comparison study, but to use these chapters to understand the complexities of mathematics teaching and learning through a curricular lens, and to highlight the deeply contextual nature of writing, intending, and enacting mathematics curricula. By looking within and across the countries and considering the similarities and differences that exist, the broader mathematics education community may envision a research agenda to assist with understanding the many nuances inherent in developing and implementing curriculum.

REFERENCES

Cai, G., & Thompson, D. R. (2014). The enacted curriculum as a focus of research. In D. R. Thompson & Z. Usiskin (Eds.), *Enacted mathematics curriculum: A conceptual framework and research needs* (pp. 1–20). Charlotte, NC: Information Age.

Eisner, E. (1985). Five basic orientations to the curriculum. In E. Eisner (Ed.), *The educational imagination: On the design and evaluation of school programs* (pp. 61–86). New York, NY: Macmillan.

Ernest, P. (1991). *The philosophy of mathematics*. New York, NY: Routledge Falmer Press.

Ernest, P. (2014). What is mathematics, and why learn it? In P. Andrews & T. Rowlands (Eds.), *Masterclass in mathematics education: International perspectives on teaching and learning* (pp. 3–14). London, England: Bloomsbury.

Grouws, D. A., Smith, M. S., & Sztajn, P. (2004). The preparation and teaching practices of United States mathematics teachers: Grades 4 and 8. In P. Kloosterman & F. K. Lester, Jr., (Eds.), *Results and interpretations of the 1990–2000 mathematics assessments of the National Assessment of Educational Progress* (pp. 221–269). Reston, VA: National Council of Teachers of Mathematics.

Kilpatrick, J. (2003). What works? In S. L. Senk & D. R. Thompson (Eds.), *Standards-based school mathematics curricula: What are they? What do students learn?* (pp. 471–488). Mahwah, NJ: Lawrence Erlbaum.

Kulm, G., & Li, Y. (2009). Curriculum research to improve teaching and learning: National and cross-national studies. *ZDM: The International Journal on Mathematics Education, 41*, 709–715.

Li, Y., & Lappan, G. (Eds.). (2014). *Mathematics curriculum in school education*. Dordrecht, the Netherlands: Springer.

Remillard, J. T., & Heck, D. J. (2014a). Conceptualizing the curriculum enactment process in mathematics education. *ZDM: The International Journal on Mathematics Education, 46*(5), 705–718.

Remillard, J. T., & Heck, D. J. (2014b). Conceptualizing the enacted curriculum in mathematics education. In D. R. Thompson & Z. Usiskin (Eds.), *Enacted mathematics curriculum: A conceptual framework and research needs* (pp. 121–148). Charlotte, NC: Information Age.

Stein, M. K., Remillard, J. T., & Smith, M. S. (2007). How curriculum influences student learning. In F. K. Lester, Jr. (Ed.), *Second handbook of research on mathematics teaching and learning* (pp. 1053–98). Charlotte, NC: Information Age.

Suurtamm, C., Thompson, D. R., Kim, R. Y., Moreno, L. D., Sayac, N., Schukajlow, S.,...Vos, P. (2016). *Assessment in mathematics education: Classroom and large-scale assessment.* Dordrecht, the Netherlands: Springer.

Tarr, J., Chávez, Ó., Reys, R. E., & Reys, B. J. (2006). From the written to the enacted curricula: The intermediary role of middle school mathematics teachers in shaping students' opportunity to learn. *School Science and Mathematics, 106*(4), 191–201.

Walker, D. (1982). Curriculum theory is many things to many people. *Theory into practice, 21*(1), 62–65.

CHAPTER 2

PRIMARY SCHOOL MATHEMATICS IN THE NETHERLANDS

The Perspective of the Curriculum Documents

Marc van Zanten
Netherlands Institute for Curriculum Development
Freudenthal Institute & Freudenthal Group, Utrecht University

Marja van den Heuvel-Panhuizen
Freudenthal Institute & Freudenthal Group, Utrecht University
Nord University, Norway

In the Netherlands, the school system consists of three stages: primary education; secondary education; and higher education (see Figure 2.1). Primary school is for students in the age range from 4 to 12 years and starts with two kindergarten grades (Grades K1 and K2), which are followed by six primary school grades (Grades 1–6). Secondary education is divided

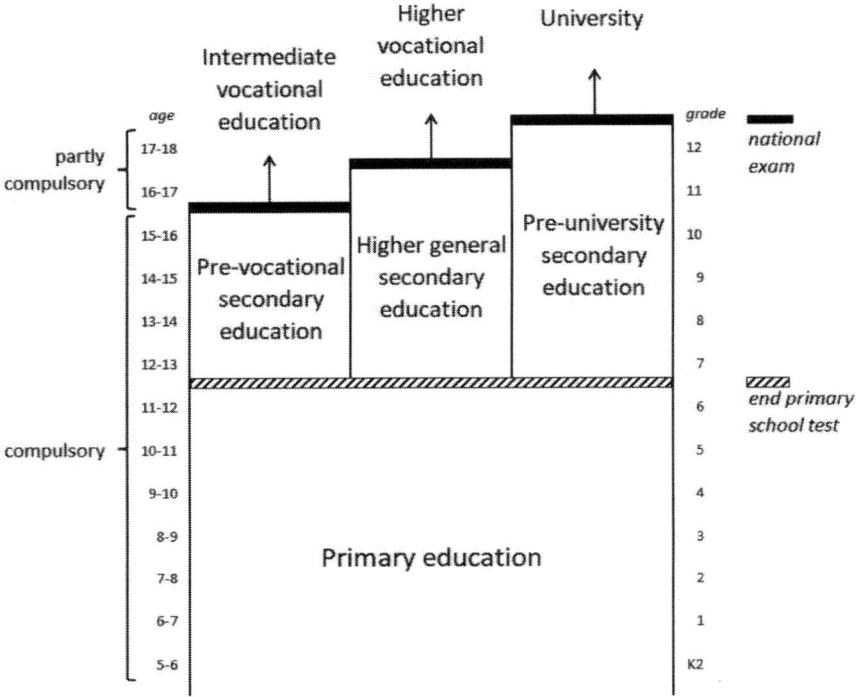

Figure 2.1 The Dutch educational system.

into three different levels with several sub-levels, and for these three levels the number of grades differs. Higher education includes vocational education and university education. Although each level of secondary education is meant to prepare students for a particular form of higher education, it is also possible for students to switch between levels. For example, a student who has attained a diploma in HAVO (higher general secondary education) can then go to the fifth and sixth grade of VWO (pre-university secondary education), and after that can go to university.

Children can go to school when they are 4 years old, but education is compulsory from the age of 5 until 16. After this age, education is partly compulsory, which means that students have to continue school until their 18th birthday or until they acquire a diploma (of HAVO, VWO or intermediate vocational education), whichever comes first.

In this chapter, we discuss the mathematics curriculum for the primary school stage. The reason for this choice is that, in the Netherlands, primary education has a longer history than secondary education in thinking about the goals to be achieved by the students. In primary education, the first goal prescriptions were released in 1993, while for secondary education they came only in 2009 and only for the first years of secondary school. For the remaining years, the curriculum is determined by the topics included in the final secondary school examinations. Moreover, the primary school mathematics curriculum is laid down in various curriculum documents, which makes it interesting to investigate how these documents together form the curriculum.

CURRICULUM DOCUMENTS FOR PRIMARY SCHOOL MATHEMATICS EDUCATION

Mathematics education starts in the kindergarten years with doing playful mathematics-related activities. In the grade years, mathematics is taught systematically in daily lessons for about five hours per week. The mathematical content that is taught in primary school is mainly defined in four types of curriculum documents:

- the legally prescribed standards;
- resources describing teaching-learning trajectories;
- textbooks; and
- assessment materials, especially compulsory tests at the end of primary school.

These documents represent different curriculum levels (e.g., Goodlad, 1979; Thijs & Van den Akker, 2009). The legally prescribed standards can be regarded as the *intended curriculum*, that is, the curriculum that describes the desired learning outcomes at a particular time in students' school career. Following Valverde, Bianchi, Wolfe, Schmidt, and Houang (2002), we consider textbooks as a separate level, the *potentially implemented curriculum*, intermediating between the intended curriculum and the implemented curriculum, which refers to the actual teaching and learning processes taking place in school (see Figure 2.2).

The teaching-learning trajectories are a mediating layer between the intended and the implemented curriculum and, therefore, belong to the potentially implemented curriculum. These trajectories sketch learning pathways through which students can achieve the standards that have been determined for the end of primary school. Although the development of these teaching-learning trajectories was initiated and financed by the Ministry of Education, they do not have a statutory status and, thus, they are not part of the formal intended curriculum. Finally, assessment materials influence the implemented curriculum because in these materials the mathematical knowledge, skills, and insights students are supposed to achieve over the school grades are operationalized.

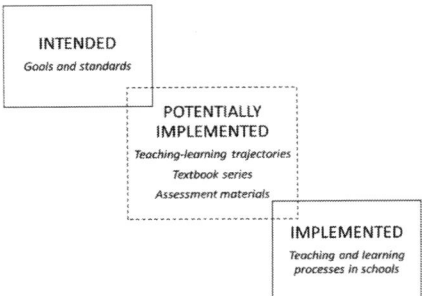

Figure 2.2 Levels of curriculum in the Netherlands (adapted version from Valverde, Bianchi, Wolfe, Schmidt, & Houang, 2002).

The aforementioned curriculum documents each have their own role in supporting mathematics education that is realized in primary school and determined by different actors, including the Ministry of Education, SLO (Netherlands Institute for Curriculum Development), CvTE (College for Tests and Examinations), Cito (National Institute for Educational Measurement), textbook authors and publishers, and developers and researchers of the Freudenthal Institute. Our aim with this chapter is to illustrate the primary mathematics curriculum in these documents and to discuss their coherence. However, to understand the role these curriculum documents play in Dutch mathematics education, we first pay attention to the constitutionally established *freedom of education* in the Netherlands.

FREEDOM OF EDUCATION

In the Netherlands, *freedom of education* implies that the government is rather restrained in being involved in how education is realized. The origin of this policy dates to the Dutch Constitution of 1848 that permitted the founding of schools based on a religious denomination (Bakker, Noordman, & Rietveld-van Wingerden, 2010). In 1917, this was followed by a law that regulated that such denominational schools from then on were to receive the same financial resources from the government as public schools (Bakker et al., 2010). A few years later, in 1920, it was decided that this regulation also applied to schools with specific pedagogical approaches (Boekholt & De Booy, 1987).

As a consequence of the restrained policy, before the first Dutch standards could be established in 1993 (Ministry of Education, 1993/1998), eight years of debate occurred around the central question of whether or not governmental prescription of goals was compatible with the freedom of education (Letschert, 1998). Since 2008, after a parliamentary inquiry of educational innovations that had taken place, the government has strived more explicitly than before to make a strict distinction between the *what* (the learning goals and content to be taught) and the *how* (the way in which this content is to be taught) of education. In that year, the parliament stated that the government only prescribes the "what," and not the "how" (Committee Parliamentary Research Education, 2008; Ministry of Education, 2008). In line with this, the government presently sees freedom of education as grounds for the founding of schools based on specific ideas about educational and didactical approaches (Education Council, 2012; Ministry of Education, 2013). Currently, the Ministry of Education (2015) is working on a law amendment for having a renewed interpretation of the freedom of education in this spirit.

As a result of the freedom of education, the Dutch government does not interfere with textbook development and there is no authority that recommends, certifies, or approves textbooks before they are put on the market. This means that there are few restrictions in developing and publishing textbooks. Schools are free to choose a textbook that they think fits most closely to their view on teaching. Regarding the compulsory student test at the end of primary education, schools have limited choice. Schools may use the test that is developed by Cito and commissioned by the government, or may use a test developed by another company but also approved by the Ministry of Education.

THE MATHEMATICS CURRICULUM AS REFLECTED IN STANDARDS

The standards for mathematics education in primary school are described in two ways. The current *Core Goals* document (Ministry of Education, 2006) describes eleven globally formulated goals, which leave much room for interpretation about what mathematics students should learn in primary school. In 2010, the Core Goals document was extended with the *Reference Framework* (Ministry of Education, 2009), which describes in more detail what students should have achieved at the end of primary school (and at the end of secondary education and at the end of intermediate vocational education).

The Core Goals for Mathematics

Figure 2.3 shows the complete list of goals for mathematics as included in the Core Goals document published in 2006. For example, for basic

> **Mathematical understanding and skills**
> 1. Students learn to use mathematical language.
> 2. Students learn to solve practical and formal mathematical problems and present their reasoning clearly.
> 3. Students learn to justify and judge solution strategies for mathematical problems.
>
> **Numbers and operations**
> 4. Students learn to understand the structure and interconnectedness of numbers, whole numbers, decimal numbers, fractions, percentages and ratios, and are able to calculate with these in practical situations.
> 5. Students learn to carry out mentally and quickly the basic operations with whole numbers at least up to 100, whereby the additions and subtractions up to 20 and the multiplication tables are known by heart.
> 6. Students learn to count and calculate by estimation.
> 7. Students learn to add, subtract, multiply and divide in clever ways.
> 8. Students learn written addition, subtraction, multiplication and division in more or less curtailed standardized ways.
> 9. Students learn to use the calculator with insight.
>
> **Measurement and geometry**
> 10. Students learn to solve simple geometry problems.
> 11. Students learn to measure and calculate with measurement units and measures related to time, money, length, perimeter, area, volume, weight, speed and temperature.

Figure 2.3 The goals in the Core Goals document for mathematics (from Ministry of Education, 2006, pp. 40–45). *Note:* All the quotations and the examples from publications published in Dutch included in this chapter have been translated into English by the authors of this chapter.

number operations, students have to learn to calculate in practical situations, and should be able to calculate mentally and in clever ways, and should be competent to carry out standardized calculation methods in a more or less curtailed way. What "practical situations" include and what these different methods imply is not specified. Regarding the number range, it is only mentioned that mental calculation should at least cover whole numbers to one hundred and that additions and subtractions up to twenty should be known by heart.

In addition to the goals, the Core Goals document also gives a so-called *characteristic of mathematics*, which describes what is valued in mathematics education. Next to the basic mathematical skills and knowledge regarding the relationships and operations that apply to numbers, measurements and structures, more overarching competencies should be valued in mathematics education, such as asking mathematical questions and problem solving. Further, it is emphasized that students should develop mathematical understanding and acquire mathematical literacy. By teachers keeping in mind students' knowledge, competencies, and interests, students "will feel challenged to carry out mathematical activity and that they will be able to do mathematics at their own level, with satisfaction and pleasure" (Ministry of

Education, 2006, p. 39). Students should also learn to respect each other's ways of thinking. Mathematics is, thus, seen as a social activity: in addition to working individually, students have to work in groups and should "learn to use explaining, formulating, notating, and giving and receiving criticism as a specific mathematical method to organize and ground their thinking and to prevent mistakes" (Ministry of Education, 2006, p. 39). A further guideline is that students should learn mathematics in the context of situations that are meaningful to them.

By including these directions in the characteristic of mathematics education, the Core Goals go, in a way, beyond prescribing just the *what* of mathematics education. They also provide a view on the learning of mathematics, which is reflected in the preamble of the Core Goals document. Although it is clearly stated that the given goals do not comment about didactics, which is in line with the freedom of education, the preamble does provide some indications about the ways in which teachers can stimulate students' development, for example, that education should be structured, interactive, and make connections to daily life (Ministry of Education, 2006, pp. 7–9).

The Reference Framework for Mathematics

The Reference Framework was developed as a result of increasing concerns about the mathematical skills of students in secondary and vocational education (Ministry of Education, 2007). This Reference Framework prescribes standards regarding the attainment targets that students should reach at specified points in their schooling, starting from the end of primary school. These attainment targets concern the domains of number, rational numbers and ratios, measurement and geometry, and data handling. For each domain, three competencies are distinguished: using mathematical language, making connections between procedures and concepts, and carrying out applications in contextual situations and bare number problems. Furthermore, for each of these competencies, three performance expectations are formulated: knowing by heart, being able to use, and understanding.

The standards are formulated for three age-related target levels (1S, 2S, 3S), and three minimum levels (1F, 2F, 3F) for students who cannot achieve the S levels. The levels 1S and 1F are meant for the end of primary school and the beginning of secondary education, in which 1S is meant for the majority of students (Expertgroep Doorlopende Leerlijnen, 2008). The 2F, 2S, 3F, and 3S levels are meant for older students. Table 2.1 shows some examples of the intended content and performance expectations for 1F and 1S in the domain of number.

TABLE 2.1 Examples of the Intended Content and Performance Expectations for 1F and 1S in the Domain of Number

Level 1F	Level 1S (Which Also Includes Level 1F)

• Translating a simple problem situation into a number sentence
• Rounding off whole numbers to round numbers
• Mental calculation: addition, subtraction, multiplication, and division "with zeroes," also with simple decimal numbers:
   30 + 50   1200 − 800
   65 × 10   3600 ÷ 100
   1000 × 2.5   0.25 × 100
• Efficient calculation (+, −, ×, ÷) using the properties of numbers and operations, with simple numbers
• Addition and subtraction (including determining the difference) with whole numbers and simple decimal numbers:
   235 + 349   1268 − 38
   €2.50 + €1.25
• Multiplication of a one-digit number with a two-digit or three-digit number:
   7 × 165
   5 hours work for €5.75 an hour
• Multiplication of a two-digit number with a two-digit number:
   35 × 67
• Division of a three-digit number with a two-digit number, with or without a remainder:
   132 ÷ 16
	• Translating a complicated problem situation into a number sentence
	• Rounding off decimal numbers to whole numbers
	• Mental calculation: addition, subtraction, multiplication, and division "with zeroes," also with more difficult numbers, including larger numbers and more complicated fractions and decimal numbers:
	   18 ÷ 100 1.8 × 1000
	• Efficient calculation with larger numbers
	• Division with a remainder or a (rounded off) decimal number:
	   122 ÷ 5

Note: From Ministry of Education (2009, pp. 23–26).

As compared with the 1F level, the 1S level generally involves handling more complex problem situations, dealing with more difficult numbers including larger numbers and complicated fractions and decimal numbers, and a higher level of understanding. For example, students have to understand the difference between a digit and a number, the importance of the number zero, and reasoning about questions like: "Does there exist a smallest fraction?" (Ministry of Education, 2009, p. 25).

The way in which the standards in the Reference Framework are formulated is more specific than in the Core Goals document. For example, in the latter document it is just stated that students have to learn to add, subtract, multiply, and divide in clever ways (see Figure 2.3). The Reference Framework is more specific about what these "clever ways" imply, namely that students should learn "efficient calculation using the properties of numbers and operations" (Ministry of Education, 2009, p. 24). In addition, compared to the Core Goals, in the Reference Framework more directions are given regarding the number range. For example, concerning multiplication, students should learn a standard procedure to multiply a three-digit number by a one-digit number, and a two-digit number by a two-digit number. Similar to the Core Goals, the Reference Framework gives no specifications or examples of efficient calculation methods or standard procedures. The same goes for descriptions as *meaningful, simple,* and *more complex* context situations. Thus, the Reference Framework, like the Core Goals, leaves much room for interpretation.

The Mathematics Curriculum as Reflected in Teaching-Learning Trajectories

In the years after 1993 when the first Core Goals were published, there was discussion about whether these end-of-primary-school standards were sufficient to ensure that these goals would be achieved (see De Wit, 1997). In particular, there was a plea for having longitudinal teaching-learning trajectories with intermediate attainment targets. In 1997, this plea for such trajectories, which were a new educational phenomenon at that time, was honored. The Ministry of Education commissioned the Freudenthal Institute to develop *TAL teaching-learning trajectories*. The acronym TAL stands for "Tussendoelen annex leerlijnen" [Intermediate attainment targets annex teaching-learning trajectories].

The first TAL trajectory (see Treffers, Van den Heuvel-Panhuizen, & Buys, 1999) was on whole-number arithmetic in the lower grades of primary school and was followed by a trajectory on whole-number arithmetic in the upper grades of primary school (see Van den Heuvel-Panhuizen, Buys, & Treffers, 2001). For the upper grades, a trajectory for rational numbers was also developed (see Van Galen et al., 2005). For the domain of measurement and geometry, a teaching-learning trajectory was developed for both the lower grades (see Van den Heuvel-Panhuizen & Buys, 2004) and for the upper grades of primary school (Gravemeijer et al., 2007).[1] Later, SLO developed online TULE[2] teaching-learning trajectories for all subjects. For mathematics, this TULE document was based on TAL. Because there are only slight differences in content between the TAL and the TULE trajectories, we confine ourselves here to a description of the TAL trajectories and, in particular, to the two on whole-number arithmetic.

In the view of the TAL developers, the term *teaching-learning trajectory*

...has three interwoven meanings: a *learning trajectory* that gives a general overview of the learning process of the students; a *teaching trajectory*, consisting of didactical indications that describe how the teaching can most effectively link up with and stimulate the learning process; and a *subject matter outline*, indicating which of the core elements of the mathematics curriculum should be taught. (Van den Heuvel-Panhuizen, 2008, p. 13)

To make the interconnectedness of learning content and didactical approach concrete, in the TAL trajectories on whole-number arithmetic there are intermediate attainment targets to serve as landmarks towards achieving the goals as included in the Core Goals document, together with *teaching frameworks*. These teaching frameworks are descriptions of the teaching-learning processes that are considered to contribute to achieving these targets. For example, regarding addition and subtraction, an intermediate attainment target says that, by the end of Grade 2, students should know how to solve addition and subtraction problems to one hundred, both in context and in a bare number format (see Table 2.2). The corresponding teaching framework indicates that, in order to reach this intermediate attainment target, the teacher should have a good understanding of the nature and function of line and group models to shift students' performance level from applying a counting strategy to a more flexible way of mental calculation and a formal way of operating with numbers.

The intermediate attainment targets and teaching frameworks form the essence of the intended teaching-learning processes. In addition, the TAL trajectories describe in full detail sequences of activities to be done, problems to be solved, strategies to be used, and the models that support these strategies. Thus, TAL provides specifications that are absent in the Core

TABLE 2.2 TAL Intermediate Attainment Target and Teaching Framework for Addition and Subtraction to One Hundred

Addition and Subtraction to 100 at the end of Grade 2	
Intermediate Attainment Target	Teaching Framework
By the end of Grade 2, the students have memorized additions and subtractions to ten and have automatized them to twenty. They should then also be able to solve addition and subtraction problems to one hundred, both in context and in a bare number format. The children may use the empty number line, write down intermediate steps, or do it entirely in their heads.	Necessary for the students to reach these attainment targets is that the teacher takes into account the different levels of the students' understanding and adapts the teaching accordingly. The teacher has to have good insight into the nature and function of line and group models. Both models facilitate the transition from the initial calculation by counting to the later, more flexible, formal operation.

Note: From Van den Heuvel-Panhuizen (2008, p. 74), based on Treffers, Van den Heuvel-Panhuizen, & Buys (1999).

Goals and the Reference Framework. For example, for standard calculation methods to one hundred (and beyond), TAL explains both the use of the *stringing strategy* (e.g., calculating 48 + 29 by doing 48 + 20 → 68 + 2 → 70 + 7 → 77) and the *splitting strategy* (e.g., calculating 48 + 29 by doing 40 + 20 = 60 and 8 + 9 = 17 followed by 60 + 17 = 77). Also, for efficient calculation methods, several varying strategies are described, such as *making use of nearby round numbers* (e.g., calculating 48 + 29 by doing 48 + 30 → 78 − 1 → 77) and *raising both terms by 1* (e.g., calculating 77 − 29 by doing 78 − 30). Furthermore, examples are given of the way in which models can be used to support specific calculation methods, such as how an empty number line can be used to solve 48 + 29 by applying a stringing strategy (see Figure 2.4a) and applying a varying strategy (see Figure 2.4b).

Another example of the specifications that TAL provides concerns two forms of written calculation procedures and their interrelatedness for the upper primary grades: whole-number-based calculation and digit-based algorithmic calculation. In the case of a whole-number-based calculation[3] of 463 + 382 (Figure 2.5a), the calculation is carried out with whole-number values working from large to small, that is from left to right (400 + 300 = 700; 60 + 80 = 140; 3 + 2 = 5; followed by 700 + 140 + 5). This calculation can also be carried out in the opposite direction working from small to large, that is from right to left (3 + 2 = 5; 60 + 80 = 140; 400 + 300 = 700; followed by 5 + 140 + 700; Figure 2.5b). By working from right to left, the procedure can be used as an introduction to digit-based algorithmic calculation[4] involving calculating with digits (3 + 2 = 5; 6 + 8 = 14, write down the 4 and carry the 1; 1 + 4 + 3 = 8; Figure 2.5c).

Similar to addition, for subtraction whole-number-based calculation and digit-based algorithmic calculation belong in TAL as common attainment targets for all students. In the case of multiplication, the most curtailed digit-based algorithmic calculation is not considered an attainment target

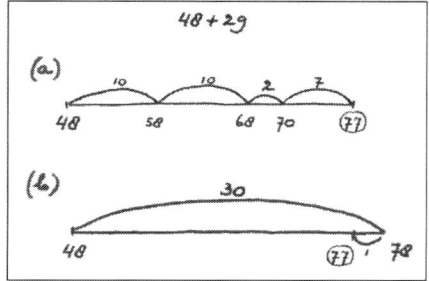

Figure 2.4a and b The use of the empty number line to support different calculation strategies for solving 48 + 29. *Source:* Van den Heuvel-Panhuizen, 2008, p. 67–68; based on Treffers, Van de Heuvel-Panhuizen, & Buys, 1999.

```
(a)   463        (b)  463         (c)   ¹463
      382 +          382 +             382 + ↓
      ───            ───               ────
      700             5                 845
      140            140 ↓
        5            ───
      ───            700
      845            ───
                     845
```

Figure 2.5a–c The addition 463 + 382 by (a) whole-number-based calculation from large to small, (b) by whole-number-based calculation from small to large, and (c) by digit-based algorithmic calculation. *Source:* Van den Heuvel-Panhuizen, 2008, p. 147; based on Treffers, Van den Heuvel-Panhuizen, & Buys, 1999.

for the lesser able students. For division, the traditional long division, the digit-based algorithmic calculation, is not considered to be an attainment target in the TAL trajectory for primary school.

Despite the detailed descriptions of the teaching-learning process for the primary school grades, the TAL trajectories are not meant to offer teachers guidance for their teaching on a day-to-day basis. The main purpose of the TAL trajectories was to bring coherence in primary school mathematics curriculum by providing a longitudinal overview of how children's mathematical understanding develops from K1 and K2 to Grade 6, and how the different stages in this development are connected and are built on each other. An example of this structure is apparent in the three levels that are distinguished in the elementary process of learning to calculate: calculating by counting (e.g., solving number problems by counting on fingers), calculating by structuring (e.g., solving number problems by using the empty number line, see Figure 2.4), and formal calculation (solving number problems by using symbolic notation). The idea is that students can solve problems at different levels, which is also recognizable in the distinction of whole-number-based calculation and digit-based algorithmic calculation. This idea reflects a concentric or spiral approach to teaching, in which a basic foundation is first laid, which later is filled with more complexity and depth. In other words, what is learned in one stage is understood in a later stage at a higher level.

Alongside the domain specific descriptions, TAL explicitly pays attention to the overarching competence of problem solving, emphasizing that students have to work on non-routine problems. For example, for the lower grades of primary school, the problem "Try to make 24 using the following randomly chosen numbers under 10: 3, 4, 7 and 8" is suggested (Van den Heuvel-Panhuizen, 2008, p. 81). In the higher grades, letter problems such

```
Find the correct digit for each letter.
The problem must match the answer.

   F  O  R  T  Y
         T  E  N
         T  E  N  +
   ─────────────────
   S  I  X  T  Y
```

Figure 2.6 Forty and ten and ten is sixty. *Source:* Van den Heuvel-Panhuizen, 2008, p. 167; see also Gardner, 1985, p. 18.

as shown in Figure 2.6, can help students to deepen their understanding of digit-based algorithmic calculation.

THE MATHEMATICS CURRICULUM AS REFLECTED IN TEXTBOOKS

Because a vast majority of Dutch primary school teachers rely heavily in their teaching on the textbook they use (Hop, 2012; Meelissen et al., 2012), mathematics textbook series have a determining role in daily teaching practice (Van Zanten & Van den Heuvel-Panhuizen, 2014). Currently, there are seven mathematics textbook series on the Dutch market, all published by independent, commercial publishers. We focus here on the four most frequently used textbook series as identified by Scheltens, Hemker, & Vermeulen (2013): *De Wereld in Getallen* (WiG; Huitema et al., 2009–2014); *Pluspunt* (PP; Van Beusekom, Fourdraine, & Van Gool, 2009–2013); *Alles Telt* (AT; Van den Bosch-Ploegh et al., 2009–2013); and *Rekenrijk* (RR; Bazen et al., 2009–2013).

All these textbook series provide materials for both students and teachers. Apart from the main books for students, the textbook series also have booklets with additional exercises and software for repetition. For Grades 1 to 6, the textbooks for students are accompanied by extensive teacher guidelines providing detailed information for each daily lesson, including directions for didactical approaches and differentiation. Moreover, these guidelines also provide, for each (sub)domain, grade overviews of the content to be addressed, and the learning goals to be achieved. For the kindergarten years, the textbook series do not have student books but only have source books for the teachers.

All textbook series offer content for numbers and operations (including whole numbers, decimal numbers, fractions, ratios and percentages, and the use of a calculator), measurement (including dealing with length, area, volume, weight, time, speed, temperature, and money), geometry (including activities that can be labeled as orienting, constructing and operating with shapes and figures), and data handling (including dealing with graphs and tables, and calculating the average of values).

Within these (sub)domains, the content and performance expectations included in the textbook series are quite similar. For example, for the domain of numbers and operations, all textbook series contain the automatizing and memorizing of addition and subtraction facts to twenty and the multiplication tables to ten; mental calculation with standard strategies and with varying strategies; estimation; written calculation in one or two standard ways (whole-number-based and digit-based-algorithmic); and making reasoned choices between mental calculation, written calculation, and using a calculator. As an example, Table 2.3 provides an overview of content and performance expectations regarding addition and subtraction in the textbook series WiG.

Although there are many similarities among the four textbook series, there are also differences, mostly related to the sequencing of the content over the grades. For example, for estimation and written calculation, the sequencing differs among the four textbook series (Table 2.4).

The performance expectations are also similar across the four textbook series. For example, they all start the automatization of adding and subtracting to 10 in Grade 1 and to 20 in Grade 2. They all also continue the process of memorizing addition and subtraction facts in Grade 3. Furthermore, all textbook series offer context situations for addition and subtraction from Grade 1 to Grade 6, first with whole numbers and later with decimal numbers in the context of money and bare decimal numbers. Another similarity is that all textbook series provide directions on how to stimulate understanding. An example is that all series explicitly offer ways to encourage students' understanding of place value, for example by using a place value chart and making references to measurement numbers (Figure 2.7).

An example of a difference in performance expectations concerns students' understanding of the relationship between whole-number-based and digit-based written calculation. For example, in WiG, AT, and RR, digit-based algorithmic written multiplication is derived from whole-number-based written multiplication, whereas in PP no relationship is explicitly made between the two forms of written multiplication. Another example concerns written addition and subtraction. RR is the only textbook series that offers whole-number-based addition and subtraction to Grade 6 (Table 2.4), which is related to what this textbook takes as a performance expectation for the lesser able students. In RR, these students may choose to apply a whole-number-based or a digit-based calculation form. Regarding multiplication, WiG and

TABLE 2.3 Overview of Content and Performance Expectations Regarding Addition and Subtraction for Grade 1 to 6 in WiG

Grade	Content and Performance Expectations
1	Addition and subtraction situations are offered for the first time. At the end of Grade 1, students have started with solving addition and subtraction to 20, both in context situations and with bare numbers, and have started automatizing splitting, adding, and subtracting with numbers to 10.
2	Students continue automatizing splitting, adding, and subtracting to 10, and start automatizing addition and subtraction to 20 and later to 100. One of the strategies students learn is making use of analogous problems ($4 + 3 \rightarrow 74 + 3$; $8 - 5 \rightarrow 48 - 5$).
3	Students continue automatizing adding and subtracting to 20. Students add and subtract to 1000, by which they make use of the decimal structure of numbers ($300 + 40$; $560 - 500$) and analogous problems to 100 ($65 + \ldots = 100 \rightarrow 165 + \ldots = 200$). All addition and subtraction problems are presented as horizontal number sentences and are calculated mentally in which the use of scrap paper and an empty number line are allowed. A start is made with using clever calculation ways ($30 + 30 \rightarrow 30 + 28$) and addition by estimation ($205 + 398 \approx$).
4	Students add and subtract to 1000 by mental calculation, also in clever ways and by estimation. Hereafter, this is extended to numbers to 10,000 and 100,000, in which students split the numbers, for example, in so many thousands, hundreds, tens, and ones. Students learn whole-number-based written addition and subtraction; after that, they learn digit-based algorithmic addition and subtraction to 1000. A start is made with digit-based algorithmic addition and subtraction with decimal numbers in the context of money.
5	Students add and subtract to 10,000 by mental calculation, also in clever ways and by estimation. Hereafter, this is extended to numbers to 1,000,000, in which students make use of decimally splitting the numbers. A start is made with adding and subtracting bare decimal numbers ($3.5 + 0.8$; $9.45 - 3.4$). Digit-based algorithmic addition and subtraction with whole numbers is done to 10,000 and with decimal numbers in the context of money up to €10,000.
6	Students add and subtract to 1,000,000 by mental calculation, also in clever ways and by estimation. Students add and subtract with decimal numbers ($2.55 + 3.5 + 102$; $7.85 - 5.4$). Digit-based algorithmic addition and subtraction with whole numbers is done to 100,000 and with decimal numbers in the context of money up to €10,000.

PP have digit-based multiplication as a goal for all students, AT has whole-number-based multiplication as a goal for lesser able students, and RR again lets lesser able students choose between whole-number-based or digit-based multiplication.

Finally, differences occur regarding the goals that textbooks set for the end of primary school. For example, the number range within which the students have to solve written multiplication problems differs among the textbooks. The textbook series WiG, AT, and RR have as a goal that students

TABLE 2.4 Sequencing of the Content Related to Addition and Subtraction (Whole Numbers and Decimal Numbers) Over the Grades in the Four Most Widely Used Dutch Textbook Series

Content	Textbook Series			
	WiG	PP	AT	RR
Addition and subtraction facts up to 20	Grades 1–3	Grades 1–3	Grades 1–3	Grades 1–3
Mental addition and subtraction in standard ways	Grades 2–6	Grades 2–6	Grades 2–6	Grades 2–6
Mental addition and subtraction in varying ways	Grades 2–6	Grades 2–6	Grades 2–6	Grades 2–6
Addition and subtraction by estimation	Grades 3–6	Grades 4–6	Grades 2–6	Grades 2–6
Whole-number-based written addition and subtraction	Grade 4	Grades 3–4	Grades 3–5	Grades 4–6
Digit-based algorithmic written addition and subtraction	Grades 4–6	Grades 4–6	Grades 3–6	Grades 4–6

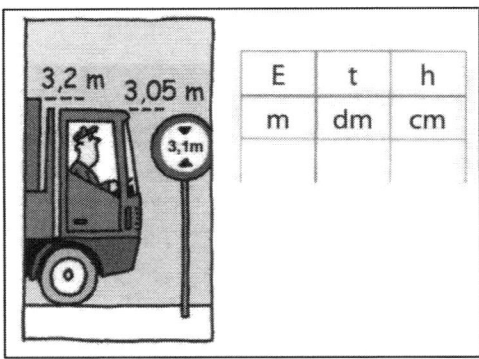

Figure 2.7 A place-value chart in WiG. *Source:* Huitema et al., 2009–2014; students' book Grade 5, p. 8. Reprinted with permission. *Note: E* = eenheden [*U* = units], *t* = tienden [*t* = tenths], *h* = honderdsten [*h* = hundredths]. In Dutch, decimal numbers have a decimal comma instead of a decimal point.

learn to multiply two-digit numbers with three-digit numbers in a digit-based algorithmic way, whereas PP does not go further than multiplying one-digit numbers with three-digit numbers and two-digit numbers with two-digit numbers.

Besides the exercises that are meant for all students, the four textbooks all provide tasks at mostly three levels. For example, WiG distinguishes so-called *one-star*, *two-stars*, and *three-stars* level tasks. Differences between these levels involve, among other things, the number range used and the complexity of the questioning. Moreover, at the one-star level, more opportunity for repetition is offered and more concrete tasks are given for a longer period of time. For example, in the final lessons in Grade 6 about multiplication by estimation, the one-star level tasks comprise estimation with decimal measurement numbers (Figure 2.8), whereas the two-star level tasks also include estimation with bare decimal numbers (Figure 2.9). The three-star level tasks require more insight, and often provide puzzle-like tasks, such as the task shown in Figure 2.10 (also from the aforementioned lesson), in which students have to use their knowledge of place value in a creative way.

What is lacking in the four textbooks is an overview of the domain-overarching competence of problem solving. This does not mean that the textbook series do not provide assignments that include problem solving.

Choose the right answer. Check your answer with a calculator.

4,5 km in 1 hour. How many kilometres in 4 hour?

4 × 4,5 km =

0,18 km

1,8 km

18 km

14,25 km in 1 hour. How many km in 5 hour?

5 × 14,25 km =

7,125 km

71,25 km

712,5 km

Figure 2.8 A Grade 6 *one-star* level task on multiplication by estimation. *Source:* Huitema et al., 2009–2014, students' book Grade 6, p. 56. Reprinted with permission.

Estimate and choose the right answer.

32 × 1,9 ≈	6,4	64	640	6400
2,9 × 40,3 ≈	1,2	12	120	1200
7 × 349,98 ≈	2,45	24,5	245	2450
98,67 × 30 ≈	20	300	200	3000
3,05 × 5,97 ≈	1,8	18	180	1800

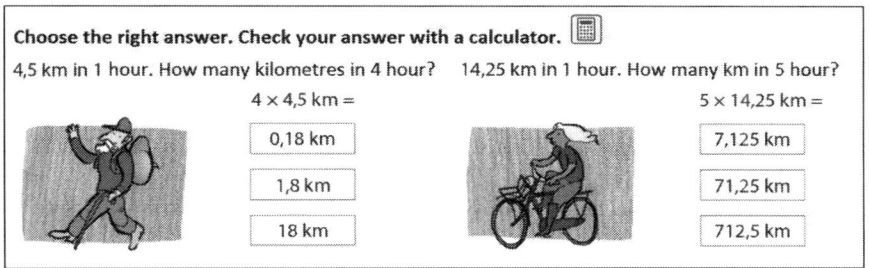

Figure 2.9 A Grade 6 *two-stars* level task on multiplication by estimation. *Source:* Huitema et al., 2009–2014, students' book Grade 6, p. 58. Reprinted with permission.

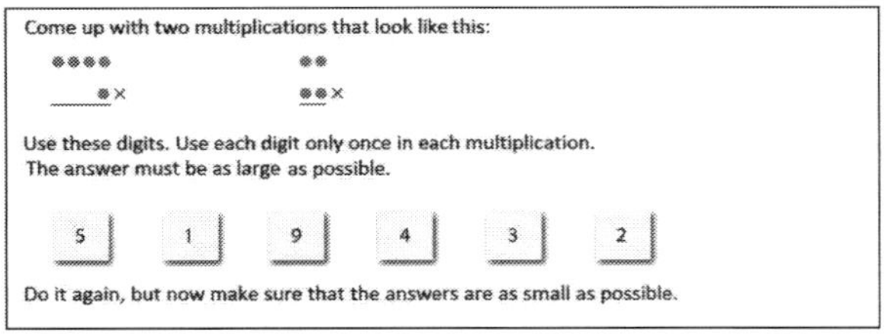

> Come up with two multiplications that look like this:
>
> ● ● ● ●   ● ●
> ____ ● ×   ● ● ×
>
> Use these digits. Use each digit only once in each multiplication.
> The answer must be as large as possible.
>
> | 5 | 1 | 9 | 4 | 3 | 2 |
>
> Do it again, but now make sure that the answers are as small as possible.

Figure 2.10 A Grade 6 *three-stars* level task on multiplication by estimation. *Source:* Huitema et al., 2009–2014, students' book Grade 6, p. 59. Reprinted with permission.

They do, but only a few assignments are included and not in a systematic way. Furthermore, problem solving tasks are mostly offered in the sections meant for the best students. In contrast, the application of mathematical knowledge and skills in solving straightforward context situations is dealt with in almost every lesson in each textbook series. Regarding another domain-overarching competence, namely using mathematical language, only AT provides an overview of mathematical words per grade.

THE MATHEMATICS CURRICULUM AS REFLECTED IN THE END OF PRIMARY SCHOOL TEST

The compulsory test at the end of primary school serves three purposes. First, the test provides objective information used in making a decision about what level of secondary education a student will attend. Second, the test results are used to know what reference level (1F or 1S) a student has mastered. Third, the test results function for the school inspectorate, next to other indicators, as a measure to assess the quality of a school. So, the end of primary school test can be considered a high-stakes test, both for students and schools.

The test that is developed by Cito and commissioned by the government is called "Centrale Eindtoets" [Central End of Primary School Test]. It is used by a majority of schools in the Netherlands (e.g., Hemker, 2016). Currently, there are several other tests developed by commercial testing companies that are approved by the government. The criteria for approval, which are also the criteria for the Central End of Primary School Test, are described in the "Toetswijzer Eindtoets PO" [Directions for End of Primary School Tests] (CvTE, 2014).

Directions for the Mathematics End of Primary School Tests

End of primary school tests must meet a number of demands with respect to validity, reliability, and content. Concerning the content, to which we confine ourselves here, end of primary school tests must cover levels 1F and 1S for all domains included in the Reference Framework (number, rational numbers and ratios, measurement and geometry, and data handling). For each domain, a minimum and maximum proportion of test items is prescribed. Also, the competencies (using mathematical language, making connections between procedures and concepts, and carrying out applications in context situations and bare number problems) named in the Reference Framework must be dealt with in an end of primary school test. The same applies to the performance expectations (knowing by heart, being able to use, and understanding).

There are also three additional specific demands. The first is that a test must contain both context problems and bare number problems, with a minimum proportion of thirty and twenty percent of all items, respectively. This demand is a direct outcome of a debate about whether mathematics education at primary school should include context situations or focus on bare number calculation. The second demand is that end of primary school tests should allow the use of scrap paper in at least eighty percent of all items, adhering to research that indicates using scrap paper was of more influence on getting a correct answer than use of a particular calculation procedure (Hickendorff, 2011). The last demand is that a test should measure whether students are able to use a calculator in a reasonable way.

The Central End of Primary School Test for Mathematics

Because a majority of schools use the Central End of Primary School Test (hereafter called the "Central Test"), we limit ourselves here to this test. The Central Test covers all the domains of the Reference Framework (Table 2.5), but not all performance expectations mentioned in the Reference Framework. This test does not (yet) contain test items assessing the ability to use a calculator, partly because this would require too many test items (CvTE, 2015a). Furthermore, the ability to make use of measurement devices is not assessed, due to the fact that the Central Test used now has a multiple-choice format.

For the Central Test, the Directions for End of Primary School Tests are extended with detailed specifications regarding the content and

TABLE 2.5 Content Included in the Central End of Primary School Test for Mathematics

Domain Mentioned in Reference Framework	Content Included in Central End of Primary School Test
Numbers	• Number sense • Operations with whole numbers and decimal numbers • Operations with fractions
Ratios	• Identifying ratios and expressing them as part-whole, fractions, percentages • Solving problems with ratios (e.g., recipes) • ...
Measurement and Geometry	• Measurement: length and circumference, area, volume, weight, time and speed, money • Geometry: shapes and figures, orientation and localization, symmetry and patterns
Data Handling	• Tables • Graphs

Note: From CvTE (2015a).

performance expectations. For example, for basic operations with whole numbers and decimal numbers, these include the following (CvTE, 2015a, pp. 53, 55):

- adding and subtracting using properties of numbers and operations, including calculation with numbers with zeroes (e.g., 4000 + 60,000; 180,000 – 2,000);
- using standard procedures for addition and subtraction with large whole numbers and decimal numbers with multiple digits;
- adding and subtracting by estimation with large whole numbers and with decimal numbers (49.95 + 128.95 + 32.35 is about 50 + 130 + 30);
- multiplying and dividing by using properties of numbers and operations, including multiplying and dividing whole numbers and decimal numbers by 10, 100, 1000 (1.8 × 100), and multiplying and dividing whole numbers by other numbers with zeroes (60 × 400; 3200 ÷ 40);
- using standard procedures for multiplication and division with large whole numbers and decimal numbers;
- interpreting the remainder of a division problem (e.g., transporting 659 children in buses; each bus can transport 45 children; 659 ÷ 45 = 14 remainder 29, so there are 15 buses needed); and
- multiplying and dividing by estimation with large whole numbers and decimal numbers (49 × 198.97 is about 50 × 200).

Because of the amount of content included in the Central Test, for language and mathematics together, it takes three mornings, including breaks, to administer the test. The 2015 version of the Central Test included 85 items for mathematics. In all items, the use of scrap paper was allowed. Figure 2.11 shows four items of the 2015 Central Test.

THE COHERENCE OF THE MATHEMATICS CURRICULUM

The coherence of a curriculum is of decisive influence on students' opportunities to learn (Schmidt, Houang, & Cogan, 2002). Curricular coherence can be considered in different ways, of which the alignment of different curriculum resources, referring to the degree in which resources agree with one another, can be seen as one of the most elementary forms (Schmidt, Wang, & McKnight, 2005). We use the term in this way, which is visualized in the Dutch *curricular spider web model* (Van den Akker, 2003; see Figure 2.12). This model illustrates the coherence of the several elements of a curriculum, but at the same time it also makes clear how vulnerable

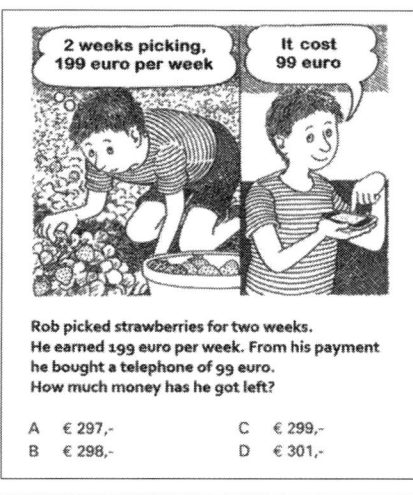

Figure 2.11 Four items on basic operations from the 2015 Central Test. *Source:* CvTE, 2015b. Reprinted with permission.

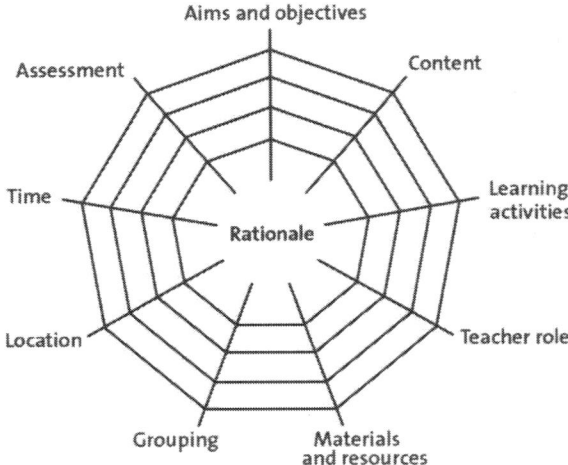

Figure 2.12 The curricular spider web. *Source:* Van den Akker, 2003.

a curriculum is. When it is pulled too hard at the ends, the spider web can break. For example, if learning materials do not fit the content to be learned, then learning goals probably will not be achieved.

The situation in which decisions regarding the curriculum are made by different actors—a government, textbook publishers, testing organizations—who each may have their own goals and visions, can be considered a threat to curricular coherence (Schmidt, Wang, & McKnight, 2005). This situation specifically applies to the Netherlands with its policy of freedom of education. Therefore, in this section, we address whether the documents that describe the intended curriculum (the Core Goals and the Reference Framework) are in alignment with each other, and whether the documents that we consider as the potentially implemented curriculum (the TAL teaching-learning trajectories, the textbook series, and the end of primary school test) correspond with the intended curriculum.

Coherence Within the Intended Curriculum

The Core Goals document and the Reference Framework give descriptions of the same content and performance expectations. All domains, content, and performance expectations included in the Core Goals document are also mentioned in the Reference Framework. The same goes for the overarching competencies of using mathematical language and problem solving, although the latter has a less prominent place in the Reference Framework than in the Core Goals document.

As noted earlier, the Reference Framework is elaborated in more detail than the Core Goals document and the Reference Framework distinguishes two levels in the attainment targets for the end of primary school. There are two other significant differences between the two documents. The first is that the Core Goals document indicates "what primary schools should be *aiming for* regarding the development of their students" (Ministry of Education, 2006, p. 1), whereas the Reference Framework describes "what students should know and be able to do regarding Dutch language and mathematics" (Ministry of Education, 2009, p. 5). Because of the latter, in 2015 the end of primary school test became mandatory, which was not previously the case. Second, although the Reference Framework contains the same content and performance expectations as the Core Goals document (CvTE, 2014), several overarching competencies emphasized in the Core Goals document are not included in the Reference Framework. This is, for example, the case for asking mathematical questions, using mathematical literacy, and giving and receiving criticism as a mathematical method. Furthermore, issues regarding attitudes mentioned in the Core Goals, such as feeling challenged and doing mathematics with satisfaction and pleasure, are also not referred to in the Reference Framework. Thus, compared to the Core Goals document, albeit the Reference Framework is more detailed in its descriptions, it is more limited with respect to mathematical attitude and overarching competence foci.

Coherence Within the Potentially Implemented Curriculum

The TAL teaching-learning trajectories, which were developed between 1996 and 2007, are based on the 1993/1998 version of the Core Goals document. Because the 2006 Core Goals document (Figure 2.3) is far more global than the 1993/1998 version was, it was expected "that the TAL teaching-learning trajectories and the included intermediate attainment targets, will play a large role in guiding decisions about mathematical content" (Van den Heuvel-Panhuizen & Wijers, 2005, p. 294). Currently, indeed, in all four most frequently used textbooks series it is explicitly stated in the accompanying teacher guidelines that the textbooks are based—next to the Core Goals and the Reference Framework—on the TAL teaching-learning trajectories (Bazen et al., 2009–2013, teacher guidelines, p. 4;[5] Huitema et al., 2009–2014, teacher guidelines, p. 2; Van Beusekom et al., 2009–2013, teacher guidelines, p. 5; Van den Bosch-Ploegh et al., 2009–2013, teacher guidelines, p. 12, p. 14). That this indeed is the case is evidenced by the corresponding ways in which content and performance expectations are aligned in the textbook series with the TAL trajectories. This is also true for the use of certain learning

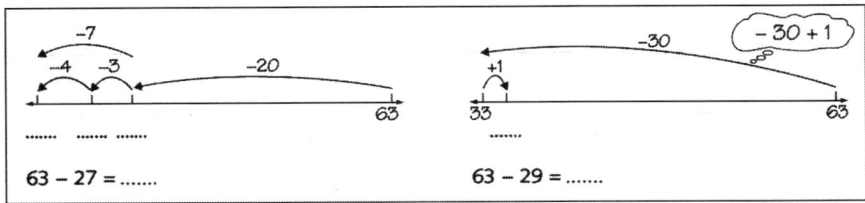

Figure 2.13 Use of the empty number line in RR. *Source:* Bazen et al., 2009–2013; students' book Grade 2, p. 45. Reprinted with permission.

facilitators as suggested by TAL, such as the empty number line, which is present in all four textbooks series (see Figure 2.13 for an example).

Despite the fact that all four textbook series have a connection with TAL, there are several differences in their elaborations of content and performance expectations (some of which were discussed in the section about textbooks) and the provision of learning facilitators, such as models. Furthermore, not everything emphasized in TAL is also present in all four textbook series. We discuss more about this in the following section.

Coherence Between the Intended and the Potentially Implemented Curriculum

The documents of the potentially implemented curriculum—the TAL teaching-learning trajectories, the four most frequently used textbook series, and the Central Test—include all the domains prescribed in the intended curriculum. They all comprise numbers and operations, ratios, measurement, geometry, and data handling. With respect to the four textbook series, analyses carried out by SLO (2012a, 2012b, 2012c, 2012d) have established that the textbooks meet the standards as described in the Core Goals document. However, it should be noted that these analyses were done very broadly and the Reference Framework was not (yet) included in these analyses.

Although the intended curriculum documents are global in nature, the potentially implemented curriculum documents provide detailed elaborations of content and performance expectations. As an example of the similarities and differences that currently exist among the curriculum documents, Table 2.6 contains a list of the ways in which written multiplication is dealt with in the Core Goals document, the Reference Framework, the TAL teaching-learning trajectories, the four textbook series, and the Central Test.

The Core Goals document and the Reference Framework prescribe that students should learn a form of written multiplication, but do not indicate what specific form (algorithmic digit-based or whole-number-based) that

TABLE 2.6 Similarities and Differences Among the Curriculum Documents for Written Multiplication

Curriculum Document	How This Document Deals With Written Multiplication
Core Goals	• Students learn written multiplication in more or less curtailed standardized ways.
Reference Framework (level 1F and 1S)	• Multiplication of a one-digit number with a two-digit or three-digit number. • Multiplication of a two-digit number with a two-digit number.
TAL Teaching-Learning Trajectory	• The most curtailed digit-based algorithmic multiplication is not considered an attainment target for lesser able students.
Textbook Series	• WiG, AT, and RR have as a goal that students learn to multiply two-digit numbers with three-digit numbers. Digit-based algorithmic multiplication is derived from whole-number-based written multiplication. • PP has as a goal that students learn to multiply one-digit numbers with three-digit numbers and two-digit numbers with two-digit numbers. No relationship is made between whole-number-based and digit-based multiplication. • WiG and PP have digit-based multiplication as a goal for all students. AT has whole-number-based multiplication as a goal for lesser able students. RR lets lesser able students choose between whole-number-based or digit-based multiplication.
Central Test	• Using standard procedures for multiplication with large whole numbers and decimal numbers.

should be. Also, the Directions for the End of Primary School Tests document do not prescribe which multiplication form should be used. The same goes for the Central Test. TAL, however, does provide an indication of the form that students should learn: the most curtailed form of digit-based multiplication is not considered an attainment target for lesser able students. The approach to written multiplication in the four textbook series varies. In agreement with TAL, in AT and RR, digit-based multiplication is not considered an attainment target for lesser able students. In WiG and PP, however, digit-based multiplication is an attainment target for all students, including the less able ones. Another difference between the textbook series is the attainment target regarding the number range in which students should be able to work. The textbook series PP has as a goal that students learn to multiply one-digit numbers with three-digit numbers and two-digit numbers with two-digit numbers, which precisely corresponds with the number range prescribed in the Reference Framework. The other textbook series aim for all students learning to multiply two-digit numbers with three-digit numbers. Finally, in WiG, AT, and RR, digit-based multiplication is derived from whole-number-based multiplication, whereas PP does not make a connection between the two forms.

Regarding the coherence between the intended curriculum and the end of primary school tests, we must say, there is a weak point. According to the Directions for the End of Primary School Tests, these tests have to "test students on their knowledge and skills regarding the Reference Framework" (CvTE, 2014, p. 17); the same document also indicates that this automatically means that the content and performance expectations of the Core Goals document are covered (CvTE, 2014). However, the latter is not necessarily true, because some overarching competencies included in the Core Goals are missing in the Reference Framework. Furthermore, some performance expectations (such as being able to use a calculator and measuring devices) are not included (yet) in the Central Test.

FINAL REMARKS

As discussed earlier, freedom of education in the Netherlands implies that there are few restrictions in developing textbooks, and that schools may choose whatever textbook series they want to use. However, because different textbooks may provide different opportunities to learn (Van Zanten & Van den Heuvel-Panhuizen, 2014) and because textbooks have a determining role for daily teaching practice in the Netherlands, we conclude this chapter with some remarks considering textbook series.

The examples provided in this chapter suggest that different elaborations within the four most frequently used textbook series fall within the boundaries of the globally described intended curriculum. However, we raise two issues.

The first issue is about the differentiated attainment targets as provided by the Reference Framework in which the levels 1F and 1S are distinguished. All four textbook series have incorporated these levels by including differentiated tasks. For example, the learning route following the *one-star* tasks in WiG is supposed to lead to mastery of the 1F level, and the route of the *two-stars* tasks should lead to the mastery of the 1S level. However, whether such differentiated learning routes within textbooks indeed lead to the mastery of the levels aimed at is not known. The fact that currently only about 45% of students at the end of primary school master the 1S level (Educational Inspectorate, 2016[6]), which is meant for a majority of the students, raises the question of whether the 1S level is well enough incorporated in the textbooks, and also how teachers deal with the differentiated routes provided by the textbooks.

The second issue concerns the domain overarching competencies, especially problem solving. Although problem solving is mentioned in both the Core Goals document and the Reference Framework, and the TAL teaching-learning trajectories explicitly emphasize the importance of it, there is only limited attention on problem solving in the four textbook series, and mainly only for the best students. This means that most students have only few opportunities to develop this mathematical competence.

Both issues—having a structure in the textbooks that clearly leads to the 1S level and offering students the opportunity to develop problem solving competencies—are definitely tasks for textbook developers to address, but to improve textbook series at this point requires that all curriculum levels be involved. Only then can the coherence of the curriculum be secured and the curriculum fulfill its role as a steering tool for high quality education.

ACKNOWLEDGMENTS

This article was partly enabled by support from the Netherlands Institute for Curriculum Development.

NOTES

1. Successively, these TAL trajectories have also been published in English (Gravemeijer et al., 2016; Van den Heuvel-Panhuizen, 2008; Van den Heuvel-Panhuizen & Buys, 2008; Van Galen et al., 2008).
2. (See http://TULE.slo.nl/). TULE stands for "Tussendoelen en leerlijnen" [Intermediate goals and teaching-learning trajectories]. Two of the three authors of TULE mathematics (Buijs, Klep, & Noteboom, 2008) were also involved in the development of TAL.
3. In the TAL teaching-learning trajectory (see Van den Heuvel-Panhuizen, 2008), this whole-number-based calculation is called *column calculation*.
4. In the TAL teaching-learning trajectory (see Van den Heuvel-Panhuizen, 2008), this digit-based algorithmic calculation is called *algorithmic calculation*.
5. In the RR guidelines, it only says "teaching-learning trajectories," but one of the authors of this textbook series confirmed that here the TAL teaching-learning trajectories are meant.
6. The same study shows that 90% of students master the 1F level at the end of primary school.

REFERENCES

Bakker, N., Noordman, J., & Rietveld-van Wingerden, M. (2010). *Vijf eeuwen opvoeden in Nederland* [Five centuries of education in the Netherlands]. (2nd Edition). Assen, the Netherlands: Van Gorcum.

Bazen, K., Bokhove, J., Borghouts, C., Buter, A., Kuipers, K., & Veltman, A. (2009–2013). *Rekenrijk 3^e editie* [Kingdom of arithmetic / Rich arithmetic]. Groningen/Houten, the Netherlands: Noordhoff Uitgevers.

Boekholt, P., & De Booy, E. (1987). *Geschiedenis van de school in Nederland*. [A history of schools in the Netherlands]. Assen, the Netherlands: Van Gorcum.
Buijs, K., Klep, J., & Noteboom, A. (2008). *TULE rekenen/wiskunde. Inhouden en activiteiten bij de kerndoelen van 2006*. [TULE Mathematics. Content and activities for the 2006 Core Goals]. Enschede, the Netherlands: Netherlands Institute for Curriculum Development.
Committee Parliamentary Research Education. (2008). *Tijd voor onderwijs*. [Time for education]. Den Haag, the Netherlands: Sdu Uitgevers.
CvTE. (2014). *Toetswijzer eindtoets PO*. [Directions for end of primary school tests]. Utrecht, the Netherlands: College voor Toetsen en Examens.
CvTE. (2015a). *Toetswijzer bij de Centrale eindtoets PO taal en rekenen*. [Directions for the Central End of primary school test for language and mathematics]. Utrecht, the Netherlands: College voor Toetsen en Examens.
CvTE. (2015b). *Central Test 2015*. Utrecht, the Netherlands: College voor Toetsen en Examens.
De Wit, C. (1997). *Over tussendoelen gesproken. Tussendoelen als component van leerlijnen*. [Talking about intermediate goals. Intermediate goals as a component of teaching-learning trajectories]. 's-Hertogenbosch, the Netherlands: KPC Onderwijs Innovatie Centrum.
Education Council. (2012). *Artikel 23 Grondwet in maatschappelijk perspectief [Article 23 of the Constitution in societal perspective]*. Den Haag, the Netherlands: Education Council of the Netherlands.
Educational Inspectorate. (2016). *Taal en rekenen aan het einde van het basisonderwijs*. [Language and mathematics at the end of primary education]. Utrecht, the Netherlands: Educational Inspectorates.
Expertgroep Doorlopende Leerlijnen. (2008). *Over de drempels met taal en rekenen*. [Crossing the thresholds with language and mathematics]. Enschede, the Netherlands: SLO.
Gardner, M. (1985). *The magic numbers of Dr. Matrix*. New York, NY: Prometeus Books.
Goodlad, J. (Ed.). (1979). *Curriculum inquiry*. New York, NY: McGraw-Hill.
Gravemeijer, K., Figueiredo, N., Feijs, E., Van Galen, F., Keijzer, R., & Munk, F. (2007). *Meten en meetkunde in de bovenbouw. Tussendoelen annex leerlijnen bovenbouw basisschool*. [Measurement and geometry in the upper grades. Teaching-learning trajectory for the upper grades of primary school]. Groningen, the Netherlands: Wolters-Noordhoff.
Gravemeijer, K., Figueiredo, N., Feijs, E., Van Galen, F., Keijzer, R., & Munk, F. (2016). *Measurement and geometry in the upper primary school*. Rotterdam, the Netherlands/Tapei, Taiwan: Sense.
Hemker, B. (2016). *Peiling van de rekenvaardigheid, de taalvaardigheid en wereldoriëntatievaardigheden in jaargroep 8 van het basisonderwijs in 2015*. [Survey on skills in mathematical, language and "world orientation" in Grade 6 in 2015]. Arnhem, the Netherlands: Cito.
Hickendorff, M. (2011). *Explanatory latent variable modeling of mathematical ability in primary school*. Leiden, the Netherlands: Leiden University (diss.).
Hop, M. (Ed.). (2012). *Balans van het reken-wiskundeonderwijs halverwege de basisschool. Periodieke peiling van het onderwijsniveau 5*. [Balance of mathematics

education halfway primary school. Periodic assessment of the education level 5]. Arnhem, the Netherlands: Cito.

Huitema, S., Erich, L., Van Hijum, R., Nillesen, C., Osinga, H., Veltman, H., & Van de Wetering, M. (2009–2014). *De Wereld in Getallen*. 's-Hertogenbosch, the Netherlands: Malmberg.

Letschert, J. (1998). *Wieden in een geheime tuin. Een studie naar kerndoelen in het Nederlandse basisonderwijs*. [Weeding in a secret garden. A study of core goals in Dutch primary education]. Enschede, the Netherlands: SLO (Netherlands Institute for Curriculum Development.

Meelissen, M. R. M., Netten, A., Drent, M., Punter, R.A., Droop, M., & Verhoeven, L. (2012). *PIRLS en TIMSS 2011. Trends in leerprestaties in Lezen, Rekenen en Natuuronderwijs*. [PIRLS and TIMSS 2011. Trends in achievement in Reading, Mathematics and Science.] Nijmegen / Enschede: Radboud University / Twente University.

Ministry of Education. (1993/1998). *Kerndoelen basisonderwijs*. [Core Goals Primary Education]. Den Haag, the Netherlands: Ministry of Education.

Ministry of Education. (2006). *Kerndoelen basisonderwijs*. [Core Goals Primary Education]. Den Haag, the Netherlands: Ministry of Education.

Ministry of Education. (2007). Reken- en taalvaardigheid en doorlopende leerlijnen. Brief aan de Tweede Kamer. [Mathematical and language skills and continual learning trajectories—Letter to the Parliament]. Den Haag, the Netherlands: Ministry of Education.

Ministry of Education. (2008). Beleidsreactie "Tijd voor onderwijs"—Brief aan de Tweede Kamer [Reaction to "Time for education"—Letter to the Parliament]. Den Haag, the Netherlands: Ministry of Education.

Ministry of Education. (2009). *Referentiekader taal en rekenen* [Reference Standards language and mathematics]. Den Haag, the Netherlands: Ministry of Education.

Ministry of Education. (2013). Artikel 23 Grondwet in maatschappelijk perspectief—Brief aan de Tweede Kamer [Article 23 of the Constitution in a societal perspective—Letter to the Parliament]. Den Haag, the Netherlands: Ministry of Education.

Ministry of Education. (2015). Meer ruimte voor nieuwe scholen: Naar een moderne interpretatie van artikel 23—Brief aan de Tweede Kamer [More room for new schools: Towards a modern interpretation of article 23—Letter to the Parliament]. Den Haag, the Netherlands: Ministry of Education.

Scheltens, F., Hemker, B., & Vermeulen, J. (2013). *Balans van het reken-wiskundeonderwijs aan het einde van de basisschool 5* [Balance of mathematics education at the end of primary school]. Arnhem, the Netherlands: Cito.

Schmidt, W., Houang, R., & Cogan, L. (2002). A coherent curriculum. The case of mathematics. *American Educator, 26(2)*, 1–17.

Schmidt, W., Wang, H., & McKnight, C. (2005). Curriculum coherence: An examination of U.S. mathematics and science content standards from an international perspective. *Journal of Curriculum Studies, 35*, 525–559.

SLO. (2012a). *Kerndoelenanalyse Alles Telt* [Core goal analysis Alles Telt]. Enschede, the Netherlands: Netherlands Institute for Curriculum Development.

SLO. (2012b). *Kerndoelenanalyse De Wereld in Getallen* [Core goal analysis De Wereld in Getallen]. Enschede, the Netherlands: Netherlands Institute for Curriculum Development.

SLO. (2012c). *Kerndoelenanalyse Pluspunt* [Core goal analysis Pluspunt]. Enschede, the Netherlands: Netherlands Institute for Curriculum Development.

SLO. (2012d). *Kerndoelenanalyse Rekenrijk* [Core goal analysis Rekenrijk]. Enschede, the Netherlands: Netherlands Institute for Curriculum Development.

Thijs, A., & Van den Akker, J. (2009). *Curriculum in development*. Enschede, the Netherlands: Netherlands Institute for Curriculum Development.

Treffers, A., Van den Heuvel-Panhuizen, M., & Buys, K. (1999). *Jonge kinderen leren rekenen. Tussendoelen Annex Leerlijnen Hele Getallen Onderbouw Basisschool* [Young children learn mathematics. A learning-teaching trajectory with intermediate attainment targets for calculation with whole numbers for the lower grades of primary school]. Groningen, the Netherlands: Wolters-Noordhoff.

Valverde, G., Bianchi, L., Wolfe, R., Schmidt, W., & Houang, R. (2002). *According to the Book. Using TIMSS to investigate the translation of policy into practice through the world of textbooks*. Dordrecht, the Netherlands: Kluwer Academic.

Van Beusekom, N., Fourdraine, A., & Van Gool, A. (Eds.). (2009–2013). *Pluspunt*. 's-Hertogenbosch, the Netherlands: Malmberg.

Van den Akker, J. (2003). Curriculum perspectives: An introduction. In J. Van den Akker, W. Kuiper, & U. Hameyer (Eds.), *Curriculum landscapes and trends*. (pp. 1–10). Dordrecht, the Netherlands: Kluwer Academic.

Van den Bosch-Ploegh, E., Van den Brom-Snijders, P., Hessing, S., Van Kraanen, H., Krol, B., Nijs-van Noort, J.,..., Vuurmans, A. (2009–2013). *Alles telt 2^e editie* [Everything counts]. Amersfoort, the Netherlands: ThiemeMeulenhoff.

Van den Heuvel-Panhuizen, M. (Ed.). (2008). *Children learn mathematics. A learning-teaching trajectory with intermediate attainment targets for calculation with whole numbers in primary school*. Rotterdam, the Netherlands/Tapei, Taiwan: Sense.

Van den Heuvel-Panhuizen, M., & Buys, K. (Eds.). (2004). *Jonge kinderen leren meten en meetkunde. Tussendoelen annex leerlijnen onderbouw basisschool* [Young children learn measurement and geometry. Teaching-learning trajectory for the lower grades of primary school]. Groningen, the Netherlands: Wolters-Noordhoff.

Van den Heuvel-Panhuizen, M., & Buys, K. (Eds.). (2008). *Young children learn measurement and geometry. A learning-teaching trajectory with intermediate attainment targets for the lower grades in primary school*. Rotterdam, the Netherlands/Tapei, Taiwan: Sense.

Van den Heuvel-Panhuizen, M., Buys, K., & Treffers, A. (Eds.). (2001). *Kinderen leren rekenen. Tussendoelen annex leerlijnen. Hele getallen bovenbouw basisschool* [Children learn mathematics. A teaching-learning trajectory for whole numbers in the upper grades of primary school]. Groningen, the Netherlands: Wolters-Noordhoff.

Van den Heuvel-Panhuizen, M. & Wijers, M. (2005). Mathematics standards and curricula in the Netherlands. *ZDM—The International Journal on Mathematics Education, 37*(4), 287–307.

Van Galen, F., Feijs, E., Figueiredo, N., Gravemeijer, K., Van Herpen, E., & Keijzer, R. (2005). *Breuken, procenten, kommagetallen en verhoudingen. Tussendoelen annex leerlijnen bovenbouw basisschool* [Fractions, percentages, decimal numbers and ratio. Teaching-learning trajectory for the upper grades of primary school]. Groningen, the Netherlands: Wolters-Noordhoff.

Van Galen, F., Feijs, E., Figueiredo, N., Gravemeijer, K., Van Herpen, E., & Keijzer, R. (2008). *Fractions, percentages, decimal numbers and ratio. A learning-teaching trajectory for Grade 4, 5 and 6*. Rotterdam, the Netherlands/Tapei, Taiwan: Sense.

Van Zanten, M., & Van den Heuvel-Panhuizen, M. (2014). Freedom of design: the multiple faces of subtraction in Dutch primary school textbooks. In Y. Li & G. Lappan (Eds.), *Mathematics curriculum in school education*. (pp. 231–259). Dordrecht, the Netherlands: Springer.

CHAPTER 3

CURRICULUM IN FRANCE
A National Frame in Transition

Ghislaine Gueudet and Lætitia Bueno-Ravel
University of Brest

Simon Modeste
University of Montpellier

Luc Trouche
Ecole Normale Supérieure de Lyon

INTRODUCTION AND HISTORICAL BACKGROUND

For the last two centuries, France has had a national curriculum elaborated by specialized committees and associated with textbooks as central for its implementation by teachers. For example, after the creation by Napoléon in 1802 of upper secondary schools ("Lycées" for students aged 11 to 18), a committee formed by three famous French mathematicians—Laplace, Monge, and Lacroix—published a sciences curriculum for these Lycées in 1803. Lacroix (1802) also wrote textbooks for the implementation of this curriculum. Some curricular reforms in France have had a strong impact on teaching beyond national borders, like the well-known reform of "Modern Maths" in the 1970s, grounding mathematics teaching on very formal approaches (Trouche, 2016a). Figure 3.1 summarizes the history of the main curriculum changes beginning in 1802. Indeed, the official curriculum changes quite often as in the recent period in 2002, 2007, 2008, and 2015–2016 for primary school. It is often said that each new government wants its own curriculum.

In this chapter, we first present an overall picture of the French mathematics curriculum, in particular since 2000; then we focus on the primary school curriculum and its recent evolutions. We follow with considerations of the controversies raised by the new curriculum (starting September 2016) for students aged 12 to 15. Next, we focus on a particular content area, algorithmics across different grades (Lagrange, 2014). Textbooks are crucial resources for teaching, influencing the implemented curriculum (Pepin & Haggarty, 2001; Valverde, Bianchi, Wolfe, Schmidt, & Houang, 2002). Deep changes in curriculum have resulted in the development of digital resources so we also focus on the development of a particular French e-textbook. In the conclusion, we synthesize the main aspects of curriculum content, design, and implementation in France.

THE NATIONAL CURRICULUM IN FRANCE: AN EVOLVING FRAMEWORK FOR ITS STRUCTURE AND CONCEPTION

In this section, we first introduce the general organization of the school system in France. We next examine the main changes in the curriculum content since 2000. Last, we discuss evolutions concerning the design mode of this curriculum.

Main Principles of the Curriculum

France is a centralized country, sharing a national curriculum for all disciplines and all class levels from Kindergarten (starting at age 3) to upper secondary school (Grade 12). An overview of the French system is presented in Table 3.1.

School is compulsory in France from age 6 to 16, but nearly all children start school at 3 years old. *Ecole maternelle* (students from ages 3 to 6) is already really a school—teachers have the same credentials as primary school teachers, and there is an official curriculum integrating mathematics through titles such as "discovering numbers and their use" or "exploring shapes and quantities." School is the same for all students from ages 6 to 15 (in *Ecole Maternelle, Ecole Primaire,* and *Collège*). The *Lycée* is organized

> **Mathematics in the French Curriculum: A Tumultuous History**
>
> The history of French mathematics curriculum is a tumultuous one that is very sensitive to scientific, social, and political tensions. For a good understanding of this history, we need to compare and contrast the *official* and the *real* curriculum. Looking at the official curriculum, the view of mathematics in France's curricula grew stronger over time, particularly in three key moments: 1802, 1902, and 2002.
>
> **1802**: Napoleon's ordinance of 19 *frimaire* of Year XI (December 10, 1802) stated: "The *Lycées* will essentially teach Latin and mathematics." Gispert (2014) notes, "In placing mathematics at the same level as Latin in the male secondary curriculum, [this ordinance] took into account the new situation following the French Revolution, in which mathematics had become *a core aspect* of an intellectual education *combining theory and practice*" (p. 230).
>
> **1902**: A new reform, following a survey launched by the French Parliament, reasserted this importance of mathematics education: "It was, for a time, the end of the monopoly on classical humanities by the Lycées, through the creation of a modern curriculum that was on par—at least in theory—with the classical curriculum. It also furthered the development of new disciplines such as the languages, sciences, and mathematics" (Gispert, 2014, p. 233).
>
> **2002**: In 1999, the French education ministry appointed a commission, the CREM (National Commission for Reflection on the Teaching of Mathematics) headed by Jean-Pierre Kahane,[a] for rethinking the teaching of mathematics for the new century. In 2002, a report of the CREM stated "La mathématique est la plus ancienne des sciences et celle dont les valeurs sont les plus permanentes"[b] (Kahane, 2002). It situated mathematics among the other sciences and underlined the necessity of connecting their teaching in combining *rigor* and *imagination*.
>
> Looking at the enacted curriculum (see e.g., Stein, Remillard, & Smith, 2007), mathematics education experienced less change than envisioned: after the ordinance of 1802, Gispert (2014, p. 230) noticed that "actually the real teaching, after this ordinance, continues to favour Latin and the classical humanities until the end of the 19th century and to separate theory and practice." It appears that two kinds of mathematics teaching existed, according to social class and schooling structure (Lycées vs. primary schools): formation of the mind in Lycées for upper social classes, training for practice in primary schools for lower social classes.
>
> Concerning the enacted curriculum in more recent years, differences with the official recommendations probably exist (e.g., concerning inquiry-based teaching, Grangeat [2011]), but we are not aware of research works proposing a comprehensive study of such differences.
>
> Questions at the heart of mathematics teaching also appeared sensitive to social and political events. Gispert (2014, p. 235) indicates, for example, "that the reform of the beginning of the 20th century—1902—was accused of being inspired by the German model of the Realschule to the detriment of the specificity of a 'French spirit' based on Latin and the classical humanities." If the place of mathematics increased during the two last centuries, it was always because of discussions full of passion, far from the view of an abstract discipline, independent of the human nature...

Figure 3.1 Mathematics in the French curriculum history.

[a] Jean-Pierre Kahane was a famous mathematician, member of the Academy of Sciences, and former president of ICMI (1983–1990). He died in 2017.

[b] "Mathematics is the most ancient science and the one whose values are the most stable" (authors' translation).

TABLE 3.1 Overview of the Educational System in France[a] as of April 2016

Age (years)	School	School Time Each Week (hours)
3 to 5–6	Ecole Maternelle (Kindergarten)	24
6 to 10–11	Ecole Primaire (primary school)	24
11 to 14–15	Collège (lower secondary)	25–29
	National Assessment: Brevet des Collèges	
15 to 17–18	Lycée Général (general upper secondary school)	28–30
	Lycée Technologique (technological upper secondary school)	
	Lycée Professionnel (vocational upper secondary school)	
	National Assessment: Baccalauréat	

[a] For more details, see http://eduscol.education.fr/cid66998/eduscol-the-portal-for-education-players.html

in three different streams: *Lycée general*, *Lycée technologique*, and *Lycée professionnel*. Within each stream, there are also different options: humanities, sciences, sciences and economics, for example, in the general stream.

In the humanities section, the main subject is philosophy (8 hours each week in Grade 12, last year of the *Lycée*); in the scientific section, the main subject is mathematics (6 hours, plus 2 hours for the students who choose the "mathematics specialty" in Grade 12). The sciences and economics section is more balanced: for grade 12 each week 5 hours of economics, 4 hours of philosophy, and 4 hours of mathematics (for the main subjects).

At the end of the *Lycée*, students stand a written examination: *Baccalauréat*. The overall success rate (general, technological, and professional) was 90.6% in 2015; around 77% of each generation obtains the baccalaureate (an aim of 80% was set in the 1980s).

Mathematics holds an important place in the curriculum for all levels and for different streams. The scientific stream at general upper secondary, labeled "S," is considered the best stream. Even at lower secondary school, mathematics is sometimes presented as a selection tool, whose usefulness resides mainly in sorting the "good" and "bad" students.

Concerning mathematics, the curriculum is presented in different official documents. The first is the "program," presenting the content and associated skills that students should master with some comments. Before 2015, all these "programs" were organized according to each grade. The new program, starting in September 2016, is written according to three-year cycles. (For a detailed analysis of the curriculum planned for Grades 1–6 and Grades 7–9, see subsequent sections.)

The program presents the content with many details. We give here a brief overview of the current Grade 10 program (started in September 2009) to provide the reader with an insight into these documents and their possible contents. The Grade 10 program comprises 10 pages, starting with an introduction stating general principles and objectives, and then giving details about three different domains (Functions, Geometry, Probability, and Statistics) and two transversal domains (Algorithmics,[1] and Reasoning and Logic). Table 3.2 presents the first line (authors' translation) of the program about Functions—the actual program table comprises eight such lines.

Other texts, called "accompanying resources," offer detailed propositions with mathematical explanations about specific subjects; for example, the accompanying resource for Algorithmics at Grade 10 is 33 pages long. These resources are written by experts (e.g., inspectors, teacher educators; we give more details later in the chapter), officially to support teachers in the design of their courses. However, teachers do not seem to use them, probably because they consider these texts as too complex. In contrast, textbook authors seem to find these "accompanying resources" very useful: almost all the examples of mathematical situations proposed in these resources are transformed into "activities" in the textbooks, after didactical transposition work (Chevallard, 1992) that makes them accessible for students.

A Process of Deep Change

The national curriculum and the associated classroom practices have deeply evolved in France from the beginning of the 21st century, in general as well as in mathematics. As mentioned in Figure 3.1, an important

TABLE 3.2 Extract of the Grade 10 Curriculum Concerning Functions

Content	Expected Skills	Comments
Functions Image, pre-image, graph	Interpret the link between two quantities with a formula. For a function defined by a graph, a table or a formula: • Identify the variable and sometimes the definition set; • Determine the image of a number; • Search for the pre-images of a number.	The functions are generally functions of one real variable, their definition set is given. Some examples of functions defined on a finite set or on integers or even functions of two variables (area as function of lengths) should be presented.

step was the report of the CREM (Kahane, 2002). One of the central objectives grounding the commission's recommendations was to bring school mathematics and "living" mathematics closer. For example, the commission recommended opening "mathematics laboratories" (Kahane, 2002, p. 268) in secondary schools, in order to "create a new image of mathematics and of its experimental aspect" (Kahane, 2002, p. 269). It also proposed a comprehensive notion of *mathematical sciences*, encompassing the mathematical practices in physics, economy, or computer science (see the commission's recommendations concerning computer science and their consequences in a later section of the chapter). The CREM emphasized the importance of reasoning and proof, stressing that "we need the alliance between imagination and reasoning present in the mathematical approach, from the formulation of statements to the proof of their consequences" (Kahane, 2002, p. 265). The orientations proposed by this commission are still very influential regarding the aims and content of mathematics teaching.

Educational policies in France are also largely influenced by the PISA (Programme for International Student Assessment) assessments and their results. The 2012 PISA tests demonstrated that, in France, students' achievements are highly correlated with their socio-economic backgrounds, and this was an important motivation for proposing a new curriculum in 2016. Concerning mathematics, the results in France are close to the Organisation for Economic Cooperation and Development (OECD) average; however, they decreased between 2003 and 2012. France was *above average* in 2003, and close to countries like Korea, Finland, and the Netherlands, whereas the 2012 results were just *on average* and on par with countries like the United Kingdom, Norway, and Denmark. In particular, the proportion of low achievers (below level 1 of PISA) increased from 5.6% in 2003 to 8.7% in 2012. The PISA tests have been cited as one of the reasons for the introduction in 2005 in the French curriculum of the *common core* (Bodin, 2008): a set of knowledge, skills, and attitudes that all students should acquire during compulsory education. The official curriculum (including this common core) has been formulated since 2005 according to knowledge, skills, and attitudes coming from the French version of the Key Competencies for Lifelong Learning (European Parliament, 2006).

The introduction of the common core and competencies was also associated with new modes of assessment; since 2009, each student has a personal competencies booklet covering all the disciplines. Two other recommended curricular evolutions linked with the common core can be considered consequences of the PISA tests: an evolution towards more individualized teaching practices taking into account each student and proposing "personal support," and the development of inquiry-based mathematics learning (Dorier & Garcia, 2013). Inquiry-based learning also aims at motivating more students for scientific studies, following the recommendations of the European Commission (Rocard et al., 2007).

The main evolutions of the curriculum since 2000 can be summarized as:

- *At a general level*—the formulation of the official curriculum and the assessment of students in terms of competencies and development of individualized practices.
- *For mathematics*—an effort to bring "living" mathematics in school, with the development of problem solving and of inquiry-based approaches (with, at the same time, an important place kept for rigorous proof at secondary school) and with an increasing importance of algorithmics.

Evolutions of the Design Mode of the Curriculum, Role of the Mathematics Education Community

France has a strong community involved in mathematics education. The teachers naturally belong to several associations, the most important being the Association of Mathematics Teachers in Public Schools (APMEP), whose website[2] and yearly conference are very popular. Most researchers in mathematics education are members of the Association for Research in Mathematics Didactics (ARDM[3]); teacher educators, a minority being also researchers, work in Schools for Teacher Education. Mathematicians working in universities can also belong to several societies, the most important being the Société Mathématique de France (SMF[4]) and the Société de Mathématiques Appliquées et Industrielles (SMAI[5]).

Within universities, "Institutes for Research on Mathematics Teaching" (IREM; Trouche, 2016a) offer opportunities to gather these different actors in research groups to study professional questions and design teaching resources. This community usually expresses its opinion regarding ongoing curriculum reforms, even if it is not in charge of the reforms. Indeed, in France several kinds of inspectors are responsible for the management of the educational system: assessing the teachers, organizing the national and regional examinations, etc. The "general inspectors" are the highest authority and are under the responsibility of the Ministry of Education; they are responsible for the text of the national curriculum. Nevertheless, groups writing this curriculum comprise different kinds of authors, most often general inspectors, local inspectors, and teachers. These groups sometimes integrate teacher educators and researchers in mathematics education, especially in the case of the primary school curriculum. Members of the IREMs (Trouche, 2016a) have been regularly involved in such groups, in particular in the CREM. They have also been involved in the commission

"Assessing the Implementation of the Grade 10 Curriculum" created in 2013. This idea of assessing the implementation and impact of a new curriculum is very recent in France.

Other significant evolutions took place in 2013 in the design process of the programs, with the creation of the "High Council of Programs" (CSP). This council, composed of members with different backgrounds (academics, members of the parliament, etc.), was created by the Ministry of Education to formulate propositions for new educational orientations and started their work in September 2014. It managed, in particular, the design of the new curriculum for primary and secondary school (from Grade 1 to Grade 9) with subgroups organized for each three-year cycle and each discipline. These groups comprised inspectors, academics, teacher educators, and teachers.

A first version of the program was published in May 2015 and a large national consultation, open to all, was organized on the ministry website for one month. Some suggestions were integrated into the final version of the program published at the end of November 2015. This means that, over the period of one year, a deep modification of the whole curriculum from Grade 1 to 9 was prepared, and should have been implemented simultaneously at all levels in September 2016.

In spite of the presence of educational researchers within the working groups, research results have not always been a central source for writing the official curriculum. Different opinions have been expressed during the discussions, and the suggestions formulated by educational researchers were not considered as more valuable than others. Nevertheless, the increasing involvement of researchers in such groups, as required by the Ministry of Education, is an important trend in the present evolutions.

FOCUS ON MATHEMATICS CURRICULUM IN PRIMARY SCHOOL

In France, primary school receives students from 6 to 10–11 years old. Until 2015, it was organized into two cycles: "Cycle 2"[6] (Grade 1 to Grade 2) and "Cycle 3" (Grade 3 to Grade 5). In September 2016, it will still be organized into two cycles, but "Cycle 2" will cover Grade 1 to Grade 3, and "Cycle 3" will cover Grade 4 to Grade 6, the first grade of lower secondary school. In this section, we present the new mathematics curriculum of primary school, which will be implemented in September 2016. Compared with the former curriculum in place since September 2008, there are many changes. We have chosen three points to illustrate the main lines of change. These points embody the aim of the CSP of improving mathematics teaching and learning. Will their implementation meet this major challenge? Further research is needed to answer this question. We present evidence of the difficulties likely to arise.

A Curriculum Reorganized Around Three-Year Cycles

The new curriculum is no longer organized by discipline, but by the common core. Taking into account the learning process of each student, even of students with specific needs, is much more manageable with a curriculum built according to three-year cycles, without giving details of what has to be done during each level of the cycle. However, such a yearly division of the curriculum is very likely to be soon available on the Internet, as teachers' main question is knowing what they have to teach in mathematics in their class level. How will the educational authorities support teachers to take into account this new way of thinking about students' learning of mathematics?

Although textbooks can support this change, most of the available updated versions as of March 2016 have continued to propose a yearly division of the mathematics curriculum. We have analyzed the new 2016 offering of seven textbook publishers. Each publisher distributes several collections of mathematics textbooks. Eleven existing collections have been updated following the new curriculum requirements, but all propose a yearly-division of the curriculum! No link seems to be made in these collections of textbooks with the last year of primary school (Grade 5, which is the second year of Cycle 3) and the first year of lower secondary school (Grade 6, which is the third and last year of Cycle 3). Furthermore, publishers' catalogs highlight interdisciplinary activities for only four collections, and activities using new technology for only three collections. Only one completely new textbook is organized at the level of the cycle: M.A.T.H. Cycle 3 (Peltier, Briand, Ngono, & Vergnes, 2016). It is structured around ten mathematics themes and a progression at the cycle level is proposed for each theme: years during which a notion of the theme is introduced, years during which a notion is studied, and years during which a notion has already been seen (consolidation). For example, for the theme "computation with decimal numbers, problems":

- The notion "multiply a decimal number by a decimal number" is introduced in Grade 4, studied in Grade 5, and considered as already been seen in Grade 6.
- The notion "column addition" is studied in Grade 4 and considered as already been seen in Grades 5 and 6.

The publisher states that this textbook is perfect as a complement to a specific grade textbook to facilitate student differentiation. It does not seem to be presented to teachers as a self-sufficient textbook.

The cycles' reform has joined the two last years of primary school with the first year of lower secondary school in the same cycle, trying to ensure continuity for students between primary and secondary school. But questions arise, such as how to develop a dialogue between primary and secondary teachers in a reasonable time frame, when primary school teachers are in charge of all the disciplines and secondary teachers are specialists of one discipline? In Grade 6, each class has 9 or 10 different teachers, one for each discipline. Will primary teachers have to find time to discuss with each of them? This question of time will be of high importance during the coming years. For example, the Cycle 2 program has 85 pages (among which only 14 concern mathematics) and the Cycle 3 program has 127 pages (among which only 17 concern mathematics). A teacher with a double level class of CE2 (Cours Elémentaire 2d year, Grade 3, last year of Cycle 2, 8–9 years old) and CM1 (Cours Moyen first year, Grade 4, first year of Cycle 3, 9–10 years old) has to read 212 pages!

A Curriculum that Takes into Account Research Findings in Mathematical Education

New programs of Cycles 2 and 3 have been elaborated by working groups comprising researchers. Some of their suggestions have been taken into account. The influence of didactical research can explain most of the changes that occurred in this new curriculum.

Concerning "space and geometry," spatial activities used to be restricted to Cycle 1, but this is no longer the case. Curriculum designers recognized the importance of spatial knowledge for the teaching of geometry according to the work of Berthelot and Salin (1999). For example, being able to use spatial references to complete a move on a map is now part of the Cycle 3 curriculum.

The topic of "Quantities and Measurement" has been greatly modified relying on the research of Chambris (2010, 2015). Relations between "quantities and measurement" and "numbers and computation" are emphasized through:

- links between teaching of the metric system and teaching of the system of place value for whole numbers instead of using mostly a conversion table (e.g., 1 m = 100 cm); in fact, the word "conversion," used before, has disappeared from the new program; and

- introduction of the "number line": a graduated straight line with numbers (Vilette, Mawart, & Rusinek, 2010) in both topics.

Concerning "numbers and calculation," the significant importance of deconstruction and reconstruction on small numbers is directly influenced by the research results of ACE[7] (Arithmetic and Comprehension in primary school), whose aim is to design a progression for numbers and calculus topics in Cycle 2. Butlen's (2007) research about mental arithmetic has also been taken into account in structuring the different types of computation in Cycle 3. Besides, following the writing of the new mathematical program concerning numbers and computation, a national conference named "consensus conference"[8] on whole numbers was organized by the National Council of Evaluation of the Scholar System (CNESCO[9]) to set up a dialogue between "experts" and members of the teaching community to establish research-based recommendations for the teaching of whole numbers.

As we can see, the new mathematics curriculum for primary school places great importance on didactical research findings and presents ambitious content related to recent research developments. Can this text be fully understood by teachers who are not specialists of mathematics and of the didactics of mathematics? The answer to this question is not clearly positive. There is a critical need for training; unfortunately, inservice teacher education has been quite reduced in France (for financial reasons in a difficult national economic context).

Using Technology in Mathematics in Primary School

Another important change in the new curriculum is the place given to the use of new technologies in mathematics classrooms. Integrating new technologies is expected starting from Grade 1 (age 6). In the former curriculum, the only mention of technological tools concerned the use of calculators in Grade 3 (age 8). Now, the use of technological tools is required in two topics: "numbers and calculation" and "space and geometry."

Concerning "numbers and calculation," calculators are introduced from Grade 1 to "calculate, estimate or verify a result." In Cycle 3, "instrumented calculation" is seen as one of the three types of calculation to practice with students; the two others are "mental arithmetic" and column method calculation. In France, the institutional demand to use calculators in primary school is not new; until now, this integration of calculators in mathematics classrooms remained low. Indeed, there is resistance to the integration of calculators (Assude, 2007; Trouche, 2016b). A number of teachers still think that using calculators will lead to poor development of students' calculation knowledge.

Concerning "space and geometry," two types of technological tools are introduced: software allowing programming of the movement of a robot or a character on a screen and dynamic geometry software (DGS). During Cycle 2, activities concerning programing the moves of a robot or a character on the screen (like with the Logo turtle) lead children to formulate simple algorithms. Such activities are related to the spatial competency "finding one's way and moving using references." In Cycle 3, an introduction to program writing is seen in "geometry" as well as the use of DGS. Using DGS in lower secondary school is quite common, but its use remains limited in primary school (Soury-Lavergne & Maschietto, 2015). Besides, many primary schools in France are still poorly equipped with computers, interactive whiteboards, and video projectors. Nevertheless, several researchers (Gueudet, Bueno-Ravel, & Poisard, 2014; Ruthven, 2012) have demonstrated that material aspects cannot be the only factor for low integration of technologies in classrooms. For example, concerning algorithmics, teachers' conceptions of mathematics (in particular the importance devoted to "rigorous proof") explain most of the failure of the first attempt to introduce algorithmics in number theory teaching in Grade 12 (age 17) in the 2000 curriculum change in secondary school (Ravel, 2003). Once again, the questions of teachers' training and teachers' resources on this subject are central. In the new textbooks, only three collections among 12 propose content in relation to the institutional requirements to use new technologies in mathematics classrooms.

FOCUS ON SOME CONTROVERSIES: "CYCLE 4" NEW CURRICULUM

The design of the new curriculum lasted for more than one year. In this section, we study three points that appeared as critical during this design process, in particular in the discussions between the CSP and the various actors of mathematics teaching: the position of mathematics among the other disciplines; the place of proving among mathematics activities; and the place of interdisciplinary activities for learning mathematics. We focus on the case of Cycle 4 (Grades 7 to 9, ages 12 to 15), which is the last cycle of the French lower secondary school.

The Position of Mathematics Among the Other Disciplines

In the global structure of the whole Cycle 4 curriculum, mathematics is mentioned at three places: in the presentation of the general features of this cycle; in the description of the contribution of each subject (mathematics, history, etc.) to the common core; and naturally in the presentation of the mathematical content to be taught. In the first component, mathematics is not explicitly mentioned. Actually, the text mentions overall general features of Cycle 4 (as "a cycle for deepening the learning") that apply to all disciplines. Regarding the second component, the discussion between CSP and the mathematics education community helped the program to evolve. At the beginning of this discussion, the domain "languages" only related to the learning of French and other languages and, actually, mathematics appeared only for its contribution to the second domain ("methods and tools for learning"). The discussion led to a more balanced view, giving to mathematics a responsibility in each domain of learning, for example:

- In the third domain (educating the person and the citizen), "mathematics as well as technical and scientific culture support the development of critical thinking and the taste for truth" (BOEN, 2015, p. 223).
- In the fourth domain (natural and technical systems), "this domain helps to initiate students to the scientific modelling, and to understand the power of mathematics." (BOEN, 2015, p. 223).

The Place of Proving in Mathematics Learning

The third component of the Cycle 4 curriculum emphasizes six main competencies for mathematics learning: searching, modelling, representing, reasoning, computing, and communicating. These competencies are clearly connected with the Key competencies defined in 2006 by the European Parliament (Table 3.3). A discussion took place about the status to be given to *proving*, seen as a critical aspect of mathematics learning. At the beginning of the discussion, proving was presented only in a restrictive way to avoid requests for students to complete formal proofs. This evidenced a kind of fear of the CSP to give too much importance to proving (perhaps due to the memory of the "Modern Math" episode), and proving was seen as restricted to geometry. The discussion led to a richer view, underlining that:

> The education to reasoning and the initiation to proving are essential objectives of the Cycle 4. The reasoning, in the heart of mathematical activity, must be based on various situations [...]. Investigative practices (trial and error, conjecture, validation, etc.) are essential and can rely both on manipulation or research on paper/pencil, on the use of digital tools (spreadsheets, DGS, etc.). It is important to provide a progression in learning to proving and not to have too many requirements on formal proof. (BOEN, 2015, p. 366)

TABLE 3.3 General Structure of Curriculum Cycle 4, Based on Three Components, From a General One to a Discipline Specific One

The Cycle 4 Curriculum, Cycle for Deepening Learning Components (1st and 2nd Common to All Disciplines, 3rd Specific for Each Discipline)	
First Component: Special Features of Cycle 4	
Presentation of the features at stake	*Disciplines mentioned for reaching this feature*
Building a new relationship with oneself and with the complexity of the world	All the disciplinary and interdisciplinary activities
Switching from one language to another	Physical, artistic, and scientific languages
Managing an abundance of information and understanding the challenges of the world	Historical dimension of knowledge
Abstracting and modelling	All the disciplines
Appropriating the great human works and developing personal creativity	Artistic and cultural activities
Being responsible and collaborating with others	Moral and civic education
Acknowledging the cultural common norms and developing a personal thinking	All the disciplines
Second Component: Contribution of Each Discipline to the Common Core	
Presentation of the 5 domains	*Place allocated to mathematics*
Languages for thinking and communicating	Languages of mathematics, of sciences and computer science
Methods and tools for learning	Mathematics, due to the necessity of memorization and to solve complex tasks
Personal and citizen education	Mathematics and scientific culture develop critical thinking and the "taste of the truth"
Natural and technical systems	Approach of the scientific modelling and first understanding of the power of mathematics
Representing the world and human activity	Scientific and technological culture; history of sciences and techniques
Third Component: Mathematics Programme for Cycle 4	
6 competencies	*Searching, Modelling, Representing, Reasoning, Computing, Communicating*
5 mathematical topics to be taught (for each topic are given: knowledge, associated competencies, and situations for constructing them)	Numbers and computation; Organizing data and function; Quantities and measurement; Space and geometry; Algorithmics and programming
8 interdisciplinary practical teaching (EPI) to be chosen by volunteer teachers	Body, health, well-being and security; Culture and artistic creation; Ecological transition and sustainable development; Information, communication, and citizenship; Languages and culture of the Antiquity; Foreign languages and cultures; Economical and professional word; Sciences, technology, and society.

Far from being restricted to mathematics, the activity of reasoning concerns all the disciplines as stated in the program: "All the disciplines are intended to underpin and broaden modes of reasoning and proving" (BOEN, 2015, p. 223). Discussions during the curriculum design process started with the intention of better situating proving within mathematics, and ended with situating reasoning and proving at the heart of learning processes.

What Should be the Place of Interdisciplinary Activities?

The new curriculum proposes a new frame for crossing disciplines: "interdisciplinary practical teaching" (Enseignements pratiques interdisciplinaires in French, EPI in the following), with six possible themes (see the third component in Table 3.3) appearing far from the usual perimeter of mathematics teaching. At first glance, the 8th theme, "science, technology and society" can appear as the only one that could be related to mathematics. The designers of the curriculum appear aware of the difficulty for mathematics teachers to contribute to the different themes grounding EPI. The curriculum underlines:

> Mathematics occupies an essential place in EPI. It provides tools for calculation and representation (using tables, diagrams, graphs), methods (based on different types of reasoning) that organize, prioritize and interpret information for various origins [...]. The variety of professions in which mathematics plays an important or essential role can be explored in the EPI. The use of media in foreign or regional language, in addition to greater exposure to other language, provides an opening to another approach to mathematics and allows students to enrol in the EPI Foreign languages and cultures. (BOEN, 2015, p. 379)

In spite of these general statements, the curriculum does not give very convincing examples of EPI giving a relevant place to mathematics. The reason is probably that the programs for each discipline have been conceived by teams of experts who are specialists in each discipline at stake. The result is that the teachers have to conceive for themselves the way to engage with their colleagues in fruitful interdisciplinary projects.

Finally, these controversies demonstrated the need for teaching resources in a time of strong evolution: evolution of the mathematical content, evolution of the teaching environment, but also evolution of the frontiers of mathematics. Kahane (2002) evokes "the mathematical sciences," and suggests envisioning mathematics as part of a network incorporating signal processing, theoretical physics, and also naturally computer science.

FOCUS ON A SPECIFIC CONTENT: ALGORITHMICS IN THE FRENCH SECONDARY SCHOOL CURRICULUM

In this section, we focus on the introduction of *algorithmics* in the mathematical curricula. By algorithmics, we refer to the branch of mathematics and computer science that is interested in the design of algorithms to solve problems and the analysis of algorithms as objects of study. (For more details, see Knuth, 2000 or Lagrange, 2014.) Such an introduction of algorithmics in connection with mathematics is specific to France and deserves to be examined here.

Computer Science and Its Relation With Mathematics in the French Curriculum: A Historical Overview

In the 1980s, after a few experiments, the teaching of computer science at upper secondary school was introduced in France for the first time with an optional teaching called "informatique des Lycées." This teaching was done by specially trained teachers, essentially mathematics teachers, and was centered on *algorithmics and programming*; the ministry invested a lot in teacher education for this option. Then, computer science as a teaching subject disappeared in the 1990s, replaced by teaching how to use computers as tools in every discipline. (See Baron & Bruillard [2011] for details about the history of teaching computer science in France.)

In the 2000s, the CREM recommended in its report to "introduce some computer science in the teaching of mathematics and in teachers' education" (Kahane, 2002, p. 44) and defended the importance of interactions between mathematics and computer science. The report addressed many arguments, summarized as follows:

- Algorithmic thinking, implicit in the teaching of mathematics, could be developed and enlightened with the instruments of algorithmics.
- Programming promotes formalized reasoning.
- Questions about effectiveness of algorithms involve mathematics.
- Data processing and digital computations are common in other disciplines.
- And finally, Computer Science transformed mathematics, bringing new points of view on objects, bringing new questions, creating new fields in mathematics that are expanding rapidly, and changing the mathematician's activity with new tools.

Just after this report, algorithmics was introduced in mathematics at Grades 11 and 12, but only for Literature and Art majors, and in optional mathematics courses (called mathematics specialty) in the last year of the economics option (in an introduction to graph theory) and the sciences option (in an introduction to number theory). Then, between 2009 and 2012 in new official programs, algorithmics was generalized in mathematics for all options of the general stream (literature, economics, sciences).

Finally, in the 2010s, computer science as a discipline came back in the upper secondary school. In 2012, a new optional teaching of computer science (called ISN, "Informatique et Sciences du Numérique") was proposed in Grade 12 for students of the scientific stream; a similar option in Grade 10 has been tried since 2015. Computer science will also be taught in Cycle 4, divided between mathematics and technology classes.

After this brief historical perspective, we examine the content of "algorithmics" in the current programs for Grades 10 to 12 in the general stream. This is followed by a discussion of "Algorithmics and Programming" in the new programs for Cycle 4.

Algorithmics in Grade 10 to 12 (Lycée, Age 15 to 18)

It is only since 2009 for Grade 10 and then since 2011 and 2012 for Grades 11 and 12 that algorithmics has been introduced in the teaching of mathematics for every option of the general stream. Exactly as in the experimentation done for Literature and Art majors, algorithmics content (just like logic content that returned to the curriculum at the same moment) has a special status in the program: it must not be taught as a course chapter (like functions, trigonometry, etc.) but integrated with the other chapters and content. The program guide asserts that "algorithms have a natural place in all the fields of mathematics" (BOEN, 2009, p. 9). Another specificity is that at all levels and in all options, the objectives are the same and given as "objective for the end of Lycée" (p. 9). This can be interpreted as a prefiguration of the organization in cycles now proposed for Grades 1 to 9. Those objectives include general competencies about "describing algorithms in natural language and symbolic languages," "realizing some algorithms" with different tools, and "interpreting more complex algorithms" (BOEN, 2009, p. 10). Some standard content in programming and algorithmics is specified as "elementary instructions (assignment, computation, input/output)" and the three classic control structures *if–then–else*, *for*, and *while* loops. Some specific algorithms or algorithmic activities are then mentioned at each level, for example, the bisection method for finding the root of a function in Grade 10 or algorithms for computing the nth element of a sequence defined by a recursive relation in Grade 11.

An accompanying resource for Grade 10, published in 2009, gives more information about the goals of this teaching of algorithmics and its place in modern mathematics. This resource has very deeply influenced the content of textbooks, and these textbooks have influenced teachers' practices. Many mathematics teachers admit that they have not been trained enough in algorithmics and have difficulties to link algorithmics with the mathematics content (Ministère de l'Éducation Nationale, 2014a).

A specific didactical transposition (Chevallard, 1992) of the concept of algorithm happened in this curriculum (Modeste, 2012). Activities in algorithmics are directed toward language activity, such as reading an algorithm, translating from natural language to a programming language, and understanding an algorithm written in a programming language. Actually, in the accompanying resource, even the concept of algorithm is defined through language: An algorithm is composed of three steps—preparation of treatment, treatment, and output of the results. The treatment is composed of instructions that are: assignation of data in variables, reading (or input) of data, sequence of instructions, and control instructions (alternative structure and repetitive structures). This construction is very close to the description of the grammar of a programming language. The resource provides a specific algorithmic language, a convention, to describe algorithms as in Figures 3.2 and 3.3.

In Figure 3.3, we notice that the "natural" language used to describe algorithms is very close to the programming language, and already includes instructions that refer to the computer (declaration of variables, printing values); the programming of the algorithm is then very close to the algorithm produced. It is a good illustration of the representation of algorithms in *Lycée*, where the notions of algorithm and program seem not very well distinguished.

Another important point is the place given to algorithmics in the experimental activity. Indeed, the role given to algorithms is often to be programmed and used to generate conjectures. For example, algorithms simulating repetitions of random experiences (rolling dice, flipping coins, random draws,...) are present at every level for illustrating properties or generating conjectures in the probability and statistics chapters. We also notice that no mathematical content or activities are proposed to deal with algorithms as objects (such as discussion about particular types of algorithms, or proof and analysis of algorithms, for example).

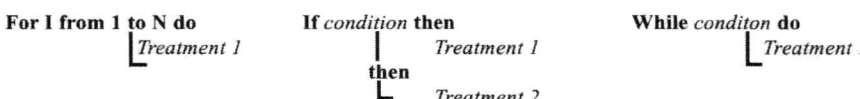

Figure 3.2 Description of the control structures in the accompanying resource (authors' translation).

Variables
    m, value of the middle of the current interval
Initialization
    a and b, bounds of the interval [a ; b]
    f, the function (reminder: the sign of f changes between a and b)
Treatment
    For i **going from** 1 **to** 50
        m **takes the value** (a+b)/2
        If f(m) **and** f(a) **have same sign then**
            a **takes the value** m
        else
            b **takes the value** m

Output
    Print a and b

Figure 3.3 An example of algorithm described in the accompanying resource, the bisection method (authors' translation).

In conclusion, the orientation of these programs is to deal with algorithms as tools for mathematics, in particular for experimental activity. This can explain the focus on language activities and the programming of most of the algorithms proposed. The final step of the didactical transposition process is curriculum implementation. Although no specific programming language is required, it seems that two types of tools for programming are currently used: the languages of calculators (mostly Casio and Texas Instruments) and a language designed by some teachers specifically for algorithmics of the *Lycée* called "Algobox."[10] We think that these languages strongly influence the activities developed by teachers.

This didactical transposition should also be compared to the curriculum of option ISN for Grade 12, where algorithmics is one of the four branches of computer science (with representation of the information, languages and programming, and material architectures). In this curriculum, algorithm is considered as a concept and defined, activities and content involve algorithms as tools and as objects, and some algorithms are introduced without being programmed.

A New Theme in Mathematics for Cycle 4: Algorithmics and Programming

In the new common core for compulsory education and the new curriculum for Cycles 3 and 4, computer science appears as a science and a technology that contribute to the development of knowledge and competencies. This is a big change for teachers who have never been trained to teach computer science; this was discussed in the process of designing the new curriculum as previously described. In Cycle 3, it will be complex to guarantee a continuity in the teaching of this new content, not only at the transition between primary school and lower secondary school (*collège*), but also at the transition with Cycle 4, which we are going to discuss below.

Concerning computer science, the most important changes take place at Cycle 4. In *collège*, as there is no independent teaching of computer science or teachers of computer science, the content has been separated between two disciplines: mathematics and technology. In mathematics, this content appears as one separated theme (of five) called "Algorithmics and Programming." It has been developed through consultation with mathematics and computer science education communities in the process previously described. The curriculum includes general content in algorithmics and programming: notions of algorithm and program, variable, instructions, parallel computing or event-driven programming (that evokes but never quotes the software Scratch[11]). It does not demand that specific algorithms or programs are taught, but gives examples of situations and activities that can be developed with the students to contribute to learning this general content. Like every discipline of the *collège*, computer science has to contribute to the interdisciplinary teaching and examples are proposed in this direction.

This new "Algorithmics and Programming" part of the curriculum raises many questions. It requires specific training for teachers, a reflection on the interactions between mathematics and computer science, and appropriate teaching resources that do not exist at this moment. It also questions the curriculum of algorithmics in the *Lycée*, as the approach of algorithmics and programming is very different in the two institutions and French students will soon be required to enter Grade 10 with competencies in computer science.

The teaching of algorithmics in secondary school in France is not stabilized yet. Through the successive changes of the curricula, it is searching for its place, between mathematics and computer science, and we anticipate that many further evolutions will happen.

IMPLEMENTATION OF THE CURRICULUM: TEXTBOOKS AND TEACHING RESOURCES

In this section we focus on textbooks, including e-textbooks in France. We first present the general situation concerning textbooks, their design and use; then we focus on a specific e-textbook designed by a teacher association.

Textbooks and Open Educational Resources in France

In France, there is no authority controlling the textbooks published; teachers are free to choose which textbook will be used by their students (we call it "the class textbook" in what follows) and free in their use of textbooks. Most of the time, the choice of the class textbook is made by the team of teachers for the grade concerned, for example, the team of mathematics teachers of Grade 10 in a given upper secondary school. At primary school and lower secondary, the class textbook is bought by the school, but at upper secondary school it is bought by the students. At secondary school, this class textbook plays a central role in the teaching-learning activity; it is used in particular to work on exercises in class and to assign homework. In previous research (e.g., Gueudet, Pepin, & Trouche, 2012; Pepin, Gueudet, & Trouche, 2013), we investigated the implementation of curriculum by teachers: selecting resources, transforming them, producing resources for students, and so on. We have documented that, in France, the textbook remains a central resource for mathematics teachers. On top of the class textbook used for the exercises, teachers use four or five other textbooks to prepare their course, to choose introductory activities, or to build assessments.

Naturally, in France like in other countries, mathematics teachers also have access to and use an abundance of teaching resources available on the Internet: lesson plans, various kinds of software, introductory activities with ready-made students' sheets, etc. The number of resources is constantly increasing; an important change occurred in 2016, linked once again with curriculum reform. The Ministry of Education selected a list of publishers to provide digital resources corresponding to the new curriculum. Teachers and students will use a range of online platforms offering these resources.

These evolutions linked with digital resources naturally also concern the textbook. All the textbooks on paper recently published in France are now associated with digital resources: A pdf version on a USB Key is offered to the teacher when his/her students buy the textbook; a website associated with the textbook offers free resources, like slides, various software files, a teacher guide, etc.; the students and the teacher can buy a "premium" digital version of the textbook, which can be annotated, complemented with external resources, etc. Because the premium digital version is very expensive, it is not much used yet. The sales figure of "premium digital textbooks" was only 1% of the textbook sales figures in 2013 (Barbat-Layani, 2013). Nevertheless, we consider that the paper textbook is now linked with an e-textbook, which is a structured system of resources (Pepin, Geuedet, Yerushalmy, Trouche, & Chazan, 2015).

These evolutions also yield changes in the design mode of textbooks (paper textbook and e-textbook). Usually in France, textbooks are written by specialists: inspectors, teacher educators, and so on. Since 2006, textbooks have also been published by an association of practicing teachers, Sésamath[12] (Gueudet, Pepin, Sabra, & Trouche, 2016). The Sésamath paper textbooks are associated with a free e-textbook, and a free complete virtual environment, LaboMEP. In the following section, we focus on the theme of functions to analyze more precisely the Sésamath Grade 10 e-textbook content.

An e-Textbook in France: Sésamath Grade 10, the Case of Functions

Sésamath is an association of mathematics teachers in France, most of whom teach at secondary school. The association was created in 2001, with the project of designing and publishing free resources for teaching and learning mathematics. The Sésamath websites receive more than 15 million hits each year. Around 20,000 teachers subscribe to the teachers' website, Sesaprof, and more than 1 million students are subscribed to LaboMEP. The use of Sésamath's online resources has been a very large scale phenomenon in France for several years.

We focus here on the Sésamath e-textbook for Grade 10. It is freely accessible online, and associated with a paper textbook that is not free, but whose price is half of the other textbooks' prices because the Sésamath association does not pay royalties.

Figure 3.4 displays a double page of the e-textbook about the variations of functions. The e-textbook contains all the content of the paper textbook,

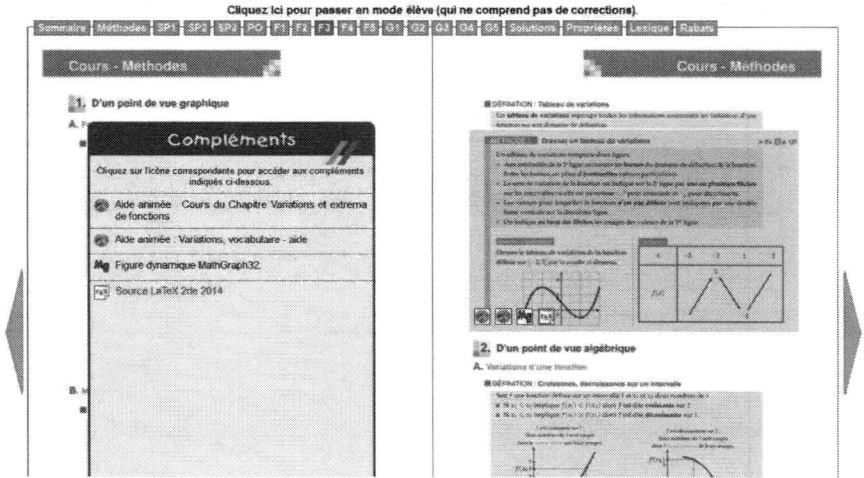

Figure 3.4 The Sésamath grade 10 e-textbook.

and possibilities of navigation in this content; on the top of Figure 3.4, we can observe an "F3" in purple color (top border strip), for Chapter 3 on functions, where we are. We can change for other chapters on geometry or statistics, and for some tools like user guides for various software. Moreover, when browsing the page with the mouse, some "complements" windows appear, offering different tools: animated helps, dynamic figures, etc. This window also offers the source file of the page: the teachers can download it to make all the modifications they wish to introduce. Concerning algorithmics, the Sésamath textbook offers exercises involving algorithms in all the chapters, including the chapters dedicated to functions. The teachers can associate these exercises with files, for example Algobox files, using LaboMEP.

The Sésamath e-textbook is directly linked with LaboMEP, which is a full virtual environment developed by the Sésamath association. The LaboMEP window (Figure 3.5) is split in three columns. In the left column, we find the resources proposed by Sésamath, in particular interactive exercises, called "Mathenpoche" (for "Math in the Pocket"). When the mouse is on a particular exercise, a small window (in blue) appears with the description of the Mathenpoche exercise. For instance, in the blue box in Figure 3.5, students are given the graph of a function and then are asked to describe it in a variety of ways. In this left column, the teacher can also have access to resources shared with his/her colleagues. This possibility of collective work is central for the members of the Sésamath association, and is not offered by other publishers yet. The Sésaprof website also offers various forums where teachers can collaborate with their colleagues and with the textbook authors. In LaboMEP, the central column is the "working column" for the teacher, where he/she can build his/her lessons, choosing

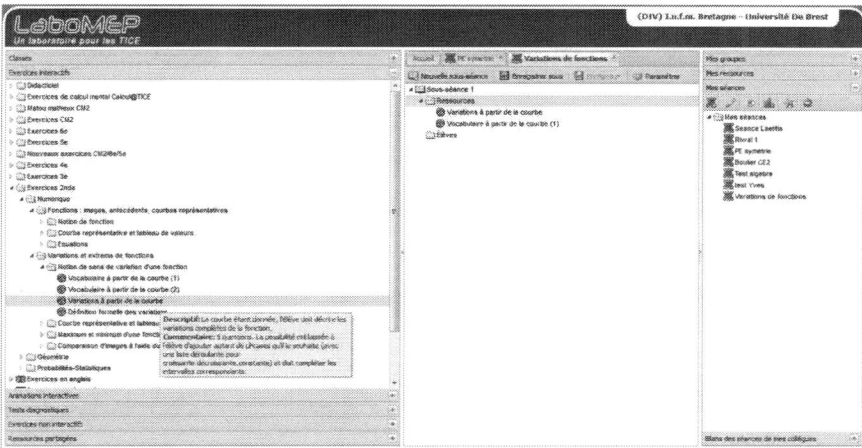

Figure 3.5 LaboMEP, the virtual learning environment of the Sésamath association.

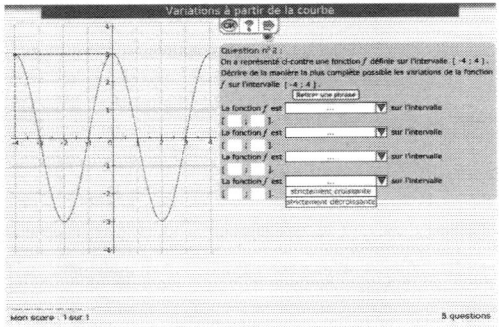

Question 2
A function f is represented on the interval [-4, 4]. Describe as completely as possible the variations of f on [-4, 4]. Students are expected to divide the interval into subintervals over which the function is strictly increasing or strictly decreasing (with those terms selected from a pull-down menu).

Figure 3.6 Mathenpoche interactive exercise, variations from the graph.

groups of students and designing for them a lesson with exercises, extracts of the e-textbook but also external (not designed by Sésamath) resources or weblinks. The right column displays all the lessons prepared by the teacher, and also gives access to the students' work on the exercises.

The "Mathenpoche" interactive exercises are the most popular resources produced by Sésamath (see Figure 3.6). Each "exercise" comprises 5 to 10 questions. The student must answer, and is allowed two attempts before losing one point, if he/she fails at the question. An animated help is associated with each exercise.

The involvement of teachers in the design of resources coordinated by Sésamath, the rich network of resources offered by this association, and the success of these resources are major phenomena that influence the implementation of the curriculum in France. It invites teachers to be involved—through platforms and other digital means—in the collective design of their teaching resources.

CONCLUSION

In this conclusion, we synthesize the central facts presented concerning the mathematics curriculum and its recent evolutions in France. This includes discussions of curriculum content and structure; curriculum design; and curriculum implementation, resources, and teachers' work.

Curriculum Content and Structure

We have described the evolution of the French curriculum over a long period as well as more recent changes. At all these different stages, rigorous proof remains an important aim of the teaching, which can be interpreted as a remnant of the "Modern Maths" reform, where "Modern Maths" now represents the traditional mathematics. Although the "rigor" philosophy is still present, the mathematics taught has also evolved and encompasses more statistics, more probability and analyses of variability, and some computer science, reflecting the evolution of contemporary mathematicians' actual work. The links between mathematics and other disciplines are emphasized; the experimental aspects and the inquiry approach are promoted. These last evolutions correspond to international trends, which also have an important influence on the French situation. In particular, the recommendations for educational policy at the European level (European Parliament, 2006) strongly influenced the choices in France, in general and also for mathematics.

Curriculum Design

Concerning the design of the official curriculum, we observed a progressive evolution towards the involvement of more actors in it. Not only the inspectors representing the institution, but also researchers in mathematics education intervene; large consultations are organized about the curriculum project, and the creation of the Superior Council for Programs constitutes a major step in the French history of curriculum design.

Curriculum Implementation, Resources, and Teachers' Work

Teachers in France do not align with a single textbook; they do significant work to design their teaching. For several years, they have had access to an abundance of resources on the web, designed not only by the educational authorities but also by individuals or associations. The use of online resources, available on different platforms, will certainly increase in the next years, in particular because the Ministry of Education has launched a call for production of online resources associated with the new 2016 curriculum. Teachers are more expected to act as designers of their mathematics courses, including now computer science and interdisciplinary aspects. At the same time, little inservice teacher education is offered. Hence, we can expect the curriculum implementation to be misaligned with the official curriculum.

Designing the curriculum and the teaching of mathematics to support learning for all students (at least in the compulsory education) remains a major challenge in France, considering poor results on national and international assessments (Keskpaik & Salles, 2013). Although France has a brilliant, internationally recognized community of mathematicians, many students encounter difficulties in mathematics, starting at primary school. This paradoxical situation led to the publication by the Ministry of Education in December 2014 of a "mathematical strategy" (Ministère de l'Education Nationale, 2014b), aiming to promote mathematics in and out of schools, and organized around three axes: new curriculum (2016); better trained teachers (this has not yet been followed by concrete decisions); changing the image of mathematics, in particular thanks to playful approaches and out-of-school activities. Thus, further analyses of the French curriculum, of its design and implementation, will certainly be needed.

NOTES

1. *Algorithmics* is mentioned throughout the chapter, as it is an important evolution in the recent French curriculum. The precise definition of this topic is given in the section focusing on it, and we invite the reader to refer to this section.
2. http://www.apmep.fr
3. http://www.ardm.eu
4. French Mathematical Society, http://smf.emath.fr/
5. Society for Applied and Industrial Mathematics, http://smai.emath.fr/
6. "Cycle 1" corresponds to preschool.
7. http://python.espe-bretagne.fr/ace/. ACE is a research project supported by the Ministry of Education.
8. http://www.cnesco.fr/fr/conference-de-consensus-numeration/
9. http://www.cnesco.fr/fr/accueil/
10. http://www.xmlmath.net/algobox/
11. Scratch (https://scratch.mit.edu) is an educational software for learning programming developed by the Massachusetts Institute of Technology and used in many countries for introducing children to programming.
12. http://www.sesamath.net

REFERENCES

Assude, T. (2007). Changements et résistances à propos de l'intégration des nouvelles technologies dans l'enseignement mathématiques au primaire [Integration of technologies at primary school: Changes and resistances]. *ISDM 29*. Retrieved from http://isdm.univ-tln.fr/PDF/isdm29/ASSUDE.pdf

Barbat-Layani, M.-A. (supervised by) (2013). *La structuration de la filière du numérique éducatif : un enjeu pédagogique et industriel* [Structuration of the digital educational resources sector: A pedagogical and industrial stake]. Retrieved from http://cache.media.education.gouv.fr/file/2013/46/0/2013-073_Numerique_educatif_271460.pdf

Baron, G.-L., & Bruillard, É. (2011). L'informatique et son enseignement dans l'enseignement scolaire général français: enjeux de pouvoirs et de savoirs [Computer science and its teaching in the French general school: Power and knowledge stakes]. In J. Lebeaume, A. Hasni, & I. Harlé (Eds.), *Recherches et expertises pour l'enseignement scientifique* [Research and expertise for science teaching] (pp. 79–90). Brussels, Belgium: De Boeck.

Berthelot, R., & Salin, M.-H. (1999). L'enseignement de l'espace à l'école primaire [Teaching space at primary school]. *Grand N, 65,* 37–59.

Bodin, A. (2008). Lecture et utilisation de PISA pour les enseignants [Reading and using PISA for teachers]. *Petit x, 78,* 53–78.

BOEN. (2009). *Programme d'enseignement de mathématiques de la classe de seconde générale et technologique* [The mathematics curriculum for general and technological grade 10]. Retrieved from http://cache.media.education.gouv.fr/file/30/52/3/programme_mathematiques_seconde_65523.pdf

BOEN. (2015). *Programmes d'enseignement du cycle des apprentissages fondamentaux (cycle 2), du cycle de consolidation (cycle 3) et du cycle des approfondissements (cycle 4)* [Curriculum for the cycle of fundamental learning (grade 1, 2 and 3), the cycle of building-up (grade 4, 5 and 6), and the cycle of deepening (grade 7, 8 and 9)]. Retrieved from http://cache.media.education.gouv.fr/file/MEN_SPE_11/67/3/2015_programmes_cycles234_4_12_ok_508673.pdf

Butlen, D. (2007). *Le calcul mental, entre sens et technique* [Mental computation, between meaning and technique]. Besançon: Presses universitaires de Franche-Comté.

Chambris, C. (2010). Relations entre grandeurs, nombres et opérations dans les mathématiques de l'école primaire au 20e siècle: Théories et écologie [Relations between quantities, numbers, and operations in the mathematics at primary school during the 20th century: Theories and ecology]. *Recherches en didactique des mathématiques, 30*(3), 317–366.

Chambris, C. (2015). Mathematical foundations for place value throughout one century of teaching in France. In X. Sun, B. Kaur, & J. Novotná (Eds.), *Proceedings of ICMI Study 23: Primary mathematics study on whole numbers* (pp. 52–59). Macau: University of Macau. Retrieved from http://www.umac.mo/fed/ICMI23/doc/Proceedings_ICMI_STUDY_23_final.pdf

Chevallard, Y. (1992). A theoretical approach to curricula. *Journal für Mathematik Didaktik, 13*(2/3), 215–230.

Dorier, J.-L, & Garcia, F. J. (2013). Challenges and opportunities for the implementation of inquiry-based learning in day-to-day teaching, *ZDM—The International Journal on Mathematics Education, 45*(6), 837–849.

European Parliament. (2006). Key competencies for lifelong learning. European Reference Framework. *Official Journal of the European Union.* Retrieved from http://eur-lex.europa.eu/LexUriServ/LexUriServ.do?uri=OJ:L:2006:394:0010:0018:en:PDF

Gispert, H. (2014). Mathematics education in France, 1900–1980. In A. Karp & G. Schubring (Eds.), *Handbook on the history of mathematics education* (pp. 229–240). New York, NY: Springer.

Grangeat, M. (2011). (Ed.). *Les démarches d'investigation dans l'enseignement scientifique Pratiques de classe, travail collectif enseignant, acquisitions des élèves* [Inquiry-based science teaching. Classroom practices, collective teachers' work, students' learning]. Lyon, France: Ecole Normale Supérieure.

Gueudet, G., Bueno-Ravel, L., & Poisard, C. (2014). Teaching mathematics with technologies at Kindergarten: Resources and orchestrations. In A. Clark-Wilson, O. Robutti, & N. Sinclair (Eds.), *The mathematics teacher in the digital era, Mathematics education in the digital era* (Vol 2., pp. 213–240). New York, NY: Springer.

Gueudet, G., Pepin, B., Sabra, H., & Trouche, L. (2016). Collective design of an e-textbook: Teachers' collective documentation. *Journal of Mathematics Teacher Education, 19*(2), 187–203.

Gueudet, G., Pepin, B., & Trouche, L. (Eds.). (2012). *From text to "lived" resources: Mathematics curriculum materials and teacher development*. New York, NY: Springer.

Kahane, J.-P. (Dir.). (2002). *L'enseignement des sciences mathématiques* [The teaching of mathematical sciences]. Paris, France: Odile Jacob.

Keskpaik, S., & Salles, F. (2013). Les élèves de 15 ans en France selon PISA 2012: Baisse des performances et augmentation des inégalités depuis 2003 [The 15 years old students in France, according to PISA 2012: Decreasing performances and increasing inequalities]. *Note d'information de la DEPP 13-31*. MENESR. Retrieved from http://cache.media.education.gouv.fr/file/2013/92/9/DEPP_NI_2013_31_eleves_15_ans_France_selon_PISA_2012_culture_mathematique_baisse_performances_augmentation_inegalites_depuis_2003_285929.pdf

Knuth, D. E. (2000). *Selected papers on analysis of algorithms*. Stanford, CA: Center for the Study of Language and Information.

Lacroix, S.-F. (1802). *Traité du calcul différentiel et du calcul intégral* (1st ed. 1797–1798) [Textbook for differential and integral calculus]. Paris, France: Duprat.

Lagrange, J.-B. (2014). Algorithmics. In S. Lerman (Ed.), *Encyclopedia of mathematics education* (pp. 32–36). Dordrecht, the Netherlands: Springer.

Ministère de l'Éducation Nationale. (2014a). L'enseignement des mathématiques, Rapport sur la mise en œuvre du programme de mathématiques en classe de seconde [Teaching mathematics, Report on the implementation of the mathematics curriculum for grade 10]. Retrieved from http://cache.media.eduscol.education.fr/file/Mathematiques/01/6/CSM-projet-rapport2013_293016.pdf

Ministère de l'Education Nationale. (2014b). *Stratégie Mathématiques* [Mathematics teaching strategy]. Retrieved from http://www.education.gouv.fr/cid84398/strategie-mathematiques.html

Modeste, S. (2012). *Enseigner l'algorithme pour quoi ? Quelles nouvelles questions pour les mathématiques? Quels apports pour l'apprentissage de la preuve ?* [Why teaching algorithm? Which new questions for mathematics? Which contributions for learning proof?]. Université de Grenoble. Retrieved from https://hal.archives-ouvertes.fr/tel-00783294v1

Peltier, M.-L., Briand, J., Ngono, B., & Vergnes, D. (2016). *MATH mathématiques cycle 3* [MATH mathematics cycle 3]. Paris, France: Hatier.

Pepin, B., Gueudet, G., Yerushalmy, M., Trouche, L., & Chazan, D. (2015). E-textbooks in/for teaching and learning mathematics: A disruptive and potentially transformative educational technology. In L. English, & D. Kirshner (Eds.), *Handbook of international research in mathematics education* (pp. 636–661). New York, NY: Taylor & Francis.

Pepin, B., Gueudet, G., & Trouche, L. (2013). Investigating textbooks as crucial interfaces between culture, policy and teacher curricular practice: Two contrasted case studies in France and Norway. *ZDM—The International Journal on Mathematics Education, 45*(5), 685–698.

Pepin, B., & Haggarty, L. (2001). Mathematics textbooks and their use in English, French and German classrooms: A way to understand teaching and learning cultures. *ZDM—The International Journal on Mathematics Education, 33*(5), 158–175.

Ravel, L. (2003). Setting a new curriculum in a classroom: Variability and space of freedom for a teacher. In M. A. Mariotti (Ed.), *Proceedings of the third conference of the European society for research in mathematics education*. Bellaria, Italy. Retrieved from http://www.mathematik.uni-dortmund.de/~erme/CERME3/Groups/TG12/TG12_Ravel_cerme3.pdf

Rocard, M., Csermely, P., Jorde, D., Lenzen, D., Walberg-Henriksson, H., & Hemmo, V. (2007). *L'enseignement scientifique aujourd'hui: Une pédagogie renouvelée pour l'avenir de l'Europe* [Science Education now: A renewed pedagogy for the future of Europe]. Bruxelles: Commission européenne. Retrieved from http://ec.europa.eu/research/science-society/document_library/pdf_06/report-rocard-on-science-education_fr.pdf

Ruthven, K. (2012). Constituting digital tools and materials as classroom resources. In G. Gueudet, B. Pepin, & L. Trouche (Eds.), *From textbooks to "lived" resources: Mathematics curriculum materials and teacher documentation* (pp. 83–103). New York, NY: Springer.

Soury-Lavergne, S., & Maschietto, M. (2015). Articulation of spatial and geometrical knowledge in problem solving with technology at primary school. *ZDM—The International Journal on Mathematics Education, 47*(3), 435–449.

Stein, M. K., Remillard, J. T., & Smith, M. S. (2007). How curriculum influences student learning. In F. K. Lester (Ed.), *Second handbook of research on mathematics teaching and learning* (pp. 319–369). Charlotte, NC: Information Age.

Trouche, L. (2016a). Didactics of mathematics: Concepts, roots, interactions and dynamics from France. In J. Monaghan, L. Trouche, & J. Borwein (Eds.), *Mathematics and tools, instruments for learning* (pp. 219–256). New York, NY: Springer.

Trouche, L. (2016b). Integrating tools as ordinary components of the curriculum in the mathematics education. In J. Monaghan, L. Trouche, & J. Borwein (Eds.), *Mathematics and tools, instruments for learning* (pp. 267–303). New York, NY: Springer.

Valverde, G., Bianchi, L. J., Wolfe, R., Schmidt, W. H., & Houang, R. T. (2002). *According to the book. Using TIMSS to investigate the translation of policy into practice through the world of textbooks*. Dordrecht, the Netherlands: Kluwer Academic.

Vilette, B., Mawart, C., & Rusinek, S. (2010). L'outil "estimateur," la ligne numérique mentale et les habiletés arithmétiques [The "estimator" tool, the mental number line and the arithmetical skills]. *Pratiques psychologiques, 16*(2), 203–214.

CHAPTER 4

MATHEMATICS CURRICULUM

The Case of Finland

Kirsti Hemmi
Åbo Akademi University and Uppsala University

Heidi Krzywacki
University of Helsinki

Anna-Maija Partanen
University of Lapland

Finland is in the middle of school reform in which the previous national core curriculum, confirmed in January 2004, is being replaced by the new curriculum, which was introduced in 2014. In this chapter, we first offer some historical background about the development of Finnish education, with a special focus on the trends in mathematics. We then describe specific features of the current educational system and policy and the context of schooling in Finland. The main focus of the chapter is to analyze the recent reform concerning mathematics curriculum, and therefore, we devote a main part of the chapter to this analysis. Finally, we discuss the current situation from an international research perspective.

A BRIEF HISTORY OF FINNISH EDUCATION

After the Second World War, Finnish society quite rapidly developed from an agrarian to a high-tech knowledge economy. At the end of the 1950s, the idea of a new education system became a focus of discussion; the function of school was to raise the competence level of *all* members of society and to develop their personality (Simola, 2002). From the 1950s onward, Finnish schools were influenced by Anglo-Saxon didactics, meaning among other things an emphasis on students' learning outcomes and personal development, and use of learner-centered methods (Lavonen & Krzywacki, 2010).

The focus in mathematics was traditionally in arithmetic and measuring during the first school grades. At the age of 10–12, students were tracked either to continue with comprehensive school or to go to a Grammar School that prepared students for further studies in upper secondary and university levels. Grammar schools had a long tradition in Finland; but only a few students had the chance to choose them, partly because of the demanding entrance examination and partly because many of them were private and had fees. Mathematics taught in these schools was theoretical and divided into algebra and geometry. Like in many European countries (cf. Törner, Potari, & Zachariades, 2014), there was an emphasis on calculus for students preparing for higher education in mathematics, science, and technology, typically to bridge the gap between school and university mathematics. There had been almost no changes in the goals and content in secondary level mathematics since the 19th century, when the first Grammar Schools were established in Finland.

The foundation of a new school system was accomplished at the end of the 1960s when comprehensive schools with no tracking were established. Educational equality, special needs education, and lifelong possibilities for learning were basic ideals of the reform. The realization of the new form of comprehensive school was carried out starting in Lapland in 1972, and moving southward to the capital area where the new comprehensive school started in 1977. At the same time, all teacher education was moved from teacher seminars to universities. Since then, both *subject teachers* who teach mathematics at lower and upper secondary school (Grades 7–12) and *class teachers* equivalent in the United States to elementary teachers, who give instruction in all subject areas (Grades 1–6), have been educated in master's level programs. The master programs comprise 300 credit points, and are offered by eight universities in Finland (Lavonen & Krzywacki, 2010).

In line with the trend in the United States and other western countries (cf. Kulm & Li, 2009), there was a short era from the 1960s to 1970s with *New Mathematics*. Finland was a part of a Nordic initiative that envisioned and developed teaching materials to support this reform in comprehensive schools (Prytz & Karlberg, 2016). This influenced also upper secondary mathematics. For example, the approach to calculus that had been quite informal with a focus on applications became more theoretical and formal according to the principles advocated in the *New Math* paradigm of the 1960s. After the *New Math* era, Finland followed a trend similar to that in many other countries—an emphasis first on basic mathematical skills, and since the 1990s also on problem-solving skills. The role of geometry, with proof and proving that had a prominent place in the curriculum of the Grammar School, was diminished in the curriculum for the comprehensive school as well as the upper secondary school (Hemmi, Lepik, & Viholainen, 2013). A more applied and informal approach to calculus was re-emphasized in upper secondary mathematics.

Still, in 1985 the national curriculum was mandated concerning the content and goals for teaching different subjects (Rokka, 2011). However, by the early 1990s, all traditional forms of control over a teacher's work, such as school inspections, a detailed national curriculum, officially approved teaching materials, weekly time schedules based on the subjects taught, and detailed class diaries had been eliminated (Simola, Rinne, Varjo, Pitkänen, & Kauko, 2009). The only remaining external control was the minimum numbers of lessons to be taught in each subject.

As Simola et al. (2009) point out, the earlier inspectorate was disliked by teachers and municipalities, and conflicted with the idea of local freedom. All these traditional means of control were replaced by local level monitoring and teacher-led processes that were to maintain the quality of teaching and learning. According to Sahlberg (2011), a typical feature of teaching and learning in Finland has been "high confidence in teachers and principals as professionals" (p. 182). The curriculum implemented in 1994 offered only very general guidelines; consistent with the general trend of decentralization, the idea was that in every school, teachers would work out a more detailed plan in every subject (Rokka, 2011). The curriculum implemented in 2004 was a step back to stronger national steering; nevertheless, the local education authorities and schools were responsible to concretize a great deal of the curriculum.

During the past four decades, the Finnish school system has succeeded in developing a high level of human capital, as well as widespread use of information and communication technologies. In addition, education and research institutions have been redesigned to foster innovation and cutting-edge research and development (Routti & Ylä-Anttila, 2006).

SPECIFIC FEATURES OF THE
EDUCATION POLICY AND SYSTEM IN FINLAND

The following section gives an overall view of the Finnish education system. This is done by first characterizing the school context, and then describing how the central ideas of the national core curriculum influence and are realized in classroom instruction.

The Context of Schooling

The Finnish education system consists of comprehensive school, post-comprehensive general (upper secondary school) and vocational education, higher education, and adult education. Comprehensive school (Grades 1–9, ages 7–16) was earlier divided into elementary (Grades 1–6) and lower secondary (Grades 7–9) school, but nowadays education policy emphasizes a common compulsory comprehensive school without division into two different school levels. Moreover, an official one-year preschool education became compulsory for all six-year-old children starting in 2015. Class teachers teach almost all subjects in elementary school in Grades 1–6, including mathematics. Subject teachers teach typically two or three school subjects each in lower (Grades 7–9) and in general upper secondary school.

In 2015, there were 2,517 comprehensive schools and 546,100 comprehensive school students educated by 39,041 teachers. About 50% of the students continue their studies in upper secondary school (Koulutuksen tilastollinen vuosikirja, 2014). A typical school has fewer than 300 students, with class sizes ranging from 15 to 30 students. Thus, schools often forge close educational communities of teachers and pupils, including parental support and involvement.

The national core curriculum, determined by the Finnish National Board of Education (FNBE), describes principles and outlines of school education in Finland; the general objectives of education as well as the objectives and core content of each school subject; principles of pupil assessment, special-needs education, pupil welfare, and educational guidance; principles of a good learning environment; and teaching methods as well as the concept of learning (FNBE, 2016). The specific objectives of instruction and content in comprehensive school mathematics are defined separately for Grades 1–2, Grades 3–6, and Grades 7–9. The national core curriculum also includes descriptions of good performance at the end of Grades 6 and 9 (with a grade "good" or 8 on a grading scale of 4–10) in all common subjects to provide teachers a tool and outline for school assessment (FNBE, 2016).

Basic education is followed typically by either general upper secondary education or vocational education (2–4 years). A combination of both fields of study is also an option in some cases. The general upper secondary education syllabus comprises a minimum of 75 courses in various subjects. Each course includes approximately 38 lesson hours and is taught in 6 to 8 week periods according to the schedule of the school. Students register for courses according to their interests, and the courses are divided into compulsory courses, optional national specialization courses, and optional local courses. The compulsory courses aim at giving a broad and general knowledge base to students, not forgetting civil skills, such as health education, that are also emphasized. At the end of upper secondary school, students take part in national matriculation examinations in 4 to 7 subjects. The tests in each subject are based on the compulsory courses and national optional courses.

Mathematics is a compulsory subject, and students choose either the *advanced level* or the *basic level* syllabus. The advanced level aims at giving students mathematical knowledge and skills needed either in university or vocational studies. The advanced level mathematics syllabus consists of ten compulsory courses and three national optional specialization courses. The basic level mathematics aims at giving understanding and readiness to use mathematical knowledge in everyday life situations as well as in further studies within social sciences and humanities. The basic level syllabus consists of six compulsory courses and two national optional specialization courses. The curriculum defines the conditions for changing the level of the syllabus that a student follows during their upper secondary studies.

Students can choose vocational education instead of the general upper secondary school after they have finished the comprehensive school (Grades 1–9). Even vocational education gives eligibility for participating in university entrance exams and thus, entering some universities, as students can study more theoretical subjects, even mathematics, and include such basic knowledge and skills in their study program. Students can also take a double exam in which they combine the vocational education with the more theoretical courses of upper secondary school.

From the National Core Curriculum to Classroom Instruction

The Finnish educational system has aimed at sustainable education by ensuring that different parts of the system are interconnected, while at the same time keeping it open to transformations (Niemi & Multisilta, 2014). Approximately every 10 years an extensive curriculum revision process is undertaken and a new curriculum is drawn up by the FNBE. The national core curriculum is prepared by working groups consisting of invited experts representing educational officials, researchers, and teachers. The preparation, led by the FNBE, is carried out in several phases starting with general structure and objectives, conceptions of learning, and learning environment. Subject-specific committees work on each school subject, like mathematics, by preparing a description of mathematics education, its content, and objectives of instruction. Since the beginning of 2000 the curriculum process has been made public in different manners, for example, through releasing information on the website of FNBE and receiving feedback at different phases during the process. Not only education providers, but also pupils and their parents are encouraged to take an active role in the process.

During the previous reform in 2000–2004, the proposed new curriculum was tested in sample schools twice during the process of reform; the teachers in these schools gave their comments and suggestions. In contrast, this time the core curriculum was prepared in a rather short time with no possibility to test it. Local education authorities have to follow the spirit and main ideas of the core curriculum, and it is their responsibility to realize the ideas at the school level. For example, there are a number of details concerning integrative instruction and school subjects that need to be outlined by local authorities. According to the new core curriculum, representatives of parents are also involved in this work, and teachers are to involve their students to plan and concretize the curriculum. It is common that municipalities carry out a curriculum, defining the objectives and contents for each grade level, with individual schools producing more concrete plans for how to enact the curriculum.

The schools and teachers themselves decide on the material and textbooks to be used. The development of mathematics textbooks, and in particular teacher guides, has had an important role in educating teachers, especially elementary teachers, concerning new ways to teach mathematics (Pehkonen, 2004). Since the 1980s, there has been a tradition of producing rich teacher guides for Grades 1–6; for example, problem solving was introduced to classroom teachers and their students through mathematics teacher guides in the 1990s. Teacher educators and researchers are often involved in the production of such material and teachers trust the quality of them (Joutsenlahti & Vainionpää, 2010; Pehkonen, 2004). The producers of curriculum materials are informed about the curriculum reform during the working process so they have the possibility of producing materials following the new national guidelines before they are implemented.

Education providers acquire all the learning materials needed, and textbooks and other materials are free for pupils. A national website (www.edu.fi), updated by the FNBE, contains information and support for teaching, such as online learning material. Still, for the past several decades, most teaching and learning materials, such as teacher guides and textbooks, are commercially produced with no national control over them. Nevertheless, we know that Finnish teachers rely to a great extent on textbooks and teacher guides, especially in mathematics (e.g., Joutsenlahti & Vainionpää, 2010).

Unlike many other countries, there is no control of students' results in the form of national examinations before the matriculation examination in upper secondary school. However, the FNBE conducts follow-up studies of basic education using a student sample, to monitor the general development of students' results, the quality and relevant distribution of special education, as well as the equity of educational qualities across different parts of Finland and across different language groups (see for example Niemi & Metsämuuronen, 2010).

THE CURRENT CURRICULUM REFORM

This section outlines some core characteristics of recent curriculum reform that was formulated in 2014 and enacted starting in 2016. We provide some general information about the underlying ideas of the curriculum and description of the features influential to the implementation before going into details related to mathematics. We conclude by discussing support for the reform.

Background

Although Finnish students have had excellent results in PISA tests (reading, science, and mathematics), young Finnish students at the lower secondary level show little interest towards these subjects and do not consider school as especially meaningful for them (Tuohilampi, Hannula, Laine, & Metsämuuronen, 2014). There is a visible trend showing that fourth-graders' interest towards mathematics has significantly diminished since the 1990s, as Finnish children's attitudes, self-confidence, and engagement are poor in comparison with other countries (Kupari, Sulkunen, Vettenranta, & Nissinen, 2012). However, in 1990, mathematics was one of the favorite subjects among Finnish fourth-graders and 40% liked mathematics very much (Kupari, 1993). Before the on-going reform, the significance of addressing attitudes and interest in the national curriculum was discussed (Kupari et al., 2012). Although the 1985 national core curriculum stressed the significance of motivational aspects in instruction, there was less focus on motivational aspects in the curricula in 1994 and in 2004 (Kupari et al., 2012).

Finland has succeeded especially well with the weakest students, but there are indications pointing to the deterioration of all Finnish students' results in mathematics (Kupari, 2013). In fact, we are now witnessing a weakening trend in students' knowledge and skills in mathematics (Niemi & Metsämuuronen, 2010). Even the results of the high-achieving students have weakened, although the number of extremely well-performing students has never been high in Finland, as international comparisons have shown.

Consistent with international discussions about current educational policies and reforms, attributes connected to the knowledge society, sustainable development, and 21st century skills, including digitalization, have been identified in Finland (cf. Sahlberg, 2011). The process of producing the new national core curriculum for basic education started in 2012 and was completed by the end of 2014. The new curriculum was adopted for students starting in the autumn of 2016. The underlying ideas of the current reform comprise so-called 21st century skills, including (a) active, student-driven knowledge creation; (b) collaboration; (c) networking; and (d) digital media competencies and literacies (Niemi & Multisilta, 2014).

Finnish educators and practitioners have stressed that an expansive use of digital technologies in education has generated the need for fresh perspectives and approaches in the development of pedagogical methods and models. There is now a strong emphasis on how to integrate technology in innovative ways that enable crossing boundaries in formal and informal learning settings (Vahtivuori-Hänninen, Halinen, Niemi, Lavonen, & Lipponen, 2014). Ketamo (2014) points out that empirical results provide evidence that students learn mathematics when gaming, for example, while teaching mathematics to virtual pets in a game. In addition, the records of user-generated behaviors could provide teachers, as well as parents, detailed information about an individual student's learning process, and thus, help in supporting individual learners.

Further, it is emphasized that ensuring all learners' involvement in school and actual learning is "a universally immense challenge, and motivational factors are crucial" (Niemi, Multisilta, Lipponen, & Vivitsou, 2014, p. x). Therefore, inspiration and joy are seen as essential in teaching and learning. An important underlying idea is how teaching and learning can be developed cooperatively with teachers, students, parents, researchers, and companies creating networking and joint aims to form an ecosystem that could support the creation of new ideas and practices. Innovative pedagogical ideas should be developed and shared and adopted by other teachers.

The new curriculum states that "each pupil has a right to grow into his or her full potential as a human being and a member of society." To achieve this, pupils need "encouragement and individual support as well as experiences of being heard and valued in the school community." The concept of learning is outlined by emphasizing the importance and willingness to act and learn in addition to the joy of learning (FNBE, 2016, pp. 15–17). The curriculum describes *transversal* (generic) *competences* as a way to meet the challenges of the future, which obviously are related to 21st century skills and integrative instruction across school subjects (FNBE, 2016, pp. 21–26). The learning objectives of the transversal competences are described as seven competence areas:

1. thinking and learning to learn;
2. cultural competence, interaction and self-expression;
3. taking care of oneself and managing daily life;
4. multiliteracy;
5. information and communication technology (ICT) competence;
6. working life competence and entrepreneurship; and
7. participation, involvement, and building a sustainable future.

FNBE calls for innovative ways in reaching the goals by local education authorities, schools, and teachers. The objectives of all subjects, including mathematics, are interconnected with the objectives of transversal competences, which are claimed to lay the ground for the overall development of an individual. These objectives will also be assessed as a part of the subject assessment so that every school subject is to enhance the development of all seven competence areas.

The importance of integration among different subjects has been addressed in the Finnish national curriculum since the 1970s. The current curriculum is the first time FNBE *demands* that all schools set the emphasis on collaborative classroom practices that include multi-disciplinary, phenomenon- and project-based studies where several teachers may work with students studying the same topic. FNBE states that all schools have to design and provide at least one such study-period per school year for all students, focused on studying phenomena or topics that are of special interest for students. Students are expected to participate in the planning process of these studies. The ethos of this new curriculum is grounded on the belief that studying will become more inspiring and meaningful if students take an active role in planning of their school work, especially the multidisciplinary study projects.

The New Mathematics Curriculum for Comprehensive School

Mathematics education is portrayed in the curriculum through describing the following themes: (a) the task of the subject and objectives of mathematics instruction; (b) key content areas in relation to the objectives of mathematics instruction; (c) objectives regarding learning environments and methods; (d) guidance, differentiation, and support; and (e) assessment. The mathematics section starts with describing the overall goals of mathematics education, which characterizes specific features of mathematics as a school subject.

Task of the Subject and Objectives of Mathematics Instruction

The national core curriculum describes the task of the subject in general and steers instruction by offering the criteria for good performance at the end of ninth grade. The curriculum implicitly guides teachers to act in certain ways in the classroom through verbalization and particular expressions. Actually, the curriculum steers instruction by describing the nature of desirable mathematics teaching. The most common expressions used in the descriptions of objectives for teaching mathematics are *support*, *guide*, *familiarize*, and *encourage*, for example, "Learning is supported by utilizing information and communication technology." Furthermore, the description of the mathematical content areas includes some steering for teachers' actions in the classroom, especially in Grades 1–6. For instance, "The pupils are guided to understand the concept of multiplication through concrete examples" and "they are guided to find and name properties that are used to classify geometric objects and plane figures" (Grades 1–2). "All arithmetic operations are practiced in versatile situations utilizing necessary tools" (Grades 3–6).

The general task of mathematics education is to develop students' logical, accurate, and creative thinking. According to the curriculum, teaching creates a basis for understanding mathematical concepts and structures in addition to students' abilities to process information and solve problems. Because mathematics is seen as a cumulative school subject, mathematics instruction proceeds systematically. Interestingly, in addition to cognitive aspects of studying and learning mathematics, the curriculum outlines objectives for developing students' attitudes and self-efficacy. During all grade levels, mathematics instruction is to strengthen students' enthusiasm and interest for mathematics, and develop their self-confidence as a mathematics learner with ability to take responsibility for their own learning. Mathematics also develops students' ability to communicate, interact, and cooperate. Mathematics education is to enhance students' ability to use and apply mathematics in various ways so that students understand the usefulness of mathematics in their own lives as well as in a broader societal perspective. The specific objectives for teaching in Grades 7–9 stress the importance of strengthening general mathematics education, deepening understanding of mathematical concepts and the relationships among them, inspiring students to discover and use mathematics in their own lives, solving problems through modeling, and encouraging students to work in a goal-oriented fashion, accurately, intently, and persistently.

The importance of some teaching methods is stressed in the description of the task of the subject. Concretizing and activating methods have a central role in teaching and learning mathematics. Further, utilizing information and communication technology is also mentioned. Students are to be offered diverse experiences as a basis for learning mathematical concepts and structures during Grades 1–2. In addition, mathematics learning takes place through different senses. During Grades 1–6, mathematics instruction aims at developing students' ability to express mathematical thoughts and solutions with the help of concrete materials, orally, in written form, through drawing, and by interpreting images and using a variety of tools. It is important to create a solid ground for understanding the concept of number and the base ten system as well as gaining elementary arithmetic skills. In Grades 3–9, a variety of problem-solving activities completed both individually and in groups, as well as comparing different solutions, comprise an essential part of mathematics teaching and learning.

Key Content Areas in Relation to the Objectives of Mathematics Instruction

The mathematical content areas comprise *thinking skills, numbers and operations, algebra, geometry and measurement, data processing, statistics and probability*, and *functions* (see Table 4.1).[1] Because the curriculum description only broadly outlines mathematical content, it allows local level actors latitude for various interpretations, especially during the first six grades. Nevertheless, there is some explicitly stated mathematical content in each grade level. The description of content areas gives an outline of what instruction should cover, and also often indicates how students are to work with the content (e.g., examine, exercise). The content is also connected to the general learning objectives.

Prescription of mathematical *thinking skills* outlines what students are to practice and master, but defines no special content or context for activities. Thus, it is the teacher's responsibility to apply appropriate content and context for acquiring such abilities. For example, finding similarities, differences, and patterns as part of logical thinking is a theme that is emphasized starting from Grades 1–2 and that deepens gradually during the school years. The pupils even familiarize themselves with proving and determining the truth value of propositions during Grades 7–9. Mathematical thinking also includes programming, which is a new school topic in Finnish school education. It applies to all students from Grade 1 up to the end of Grade 9. Understanding the basis of programming and algorithmic thinking is thought to contribute to 21st century skills. However, it is not easy to include such a new strand within a traditional school subject like mathematics. First, learning programming starts with constructing simple algorithmic instructions by using symbols in written or oral form and testing them. During Grades 3–6, the emphasis is on formulating instructions in a graphic programming environment. Not until Grades 7–9 are students

TABLE 4.1   Examples of the Central Content at Different Grade Levels in Comprehensive Schools

Content Area	Grades 1–2	Grades 3–6	Grades 7–9
Thinking skills	The pupils are provided with opportunities to find similarities, differences and regularities. The pupils compare, classify, and place objects in order and identify causal relationships.	The pupils develop their skills in finding similarities, differences, and regularities. They also improve their skills in comparing, classifying, and arranging objects, systematically searching for alternatives, and observing causal relationships and connections in mathematics.	The pupils practice activities requiring logical thinking, such as discovering rules and dependencies and presenting them accurately.... Their reasoning and argumentation skills are strengthened.... They familiarize themselves with the basics of providing proof.
	The pupils begin familiarizing themselves with the basics of programming by formulating and testing step-by-step instructions.	The pupils plan and execute programs in graphic programming environments.	The pupils deepen their algorithmic thinking. They program while learning good programming practices. The pupils use their own or ready-made computer programs as a part of learning mathematics.
Numbers and operations	Operations are performed using natural numbers.... Pupils examine the properties of numbers, such as parity, multiples, and divisions by two.... The pupils familiarize themselves with the principles of base-ten system using concrete models....	It is ensured that the pupils understand base-ten system and can deepen their understanding of it. The pupils examine and classify numbers to diversify their perception of connections between and the structure and divisibility of numbers.... All operations are practiced in versatile situations, utilizing the necessary tools....	The pupils practice basic arithmetic operations also with negative numbers. They strengthen their arithmetic skills using fractions and learn to multiply and divide by fractions.... It is ensured that the pupils understand the concept of percentages....
	The pupils learn to use the commutative and associative properties of addition.	Pupils utilize the properties of operations and the connections between them.	

(continued)

Mathematics Curriculum: The Case of Finland ■ 83

TABLE 4.1 Examples of the Central Content at Different Grade Levels in Comprehensive Schools (continued)

Content Area	Grades 1–2	Grades 3–6	Grades 7–9
Algebra (3–9)		The pupils observe the regularities of number sequences and continue a number sequence following its rule. They get to know the concept of unknown. They examine equations and solve them by reasoning and experimentation.	The pupils familiarize themselves with the concept of variable and calculating the value of a mathematical expression. They practice simplifying exponential expressions.... They solve pairs of equations graphically and algebraically. They familiarize themselves with first-degree inequalities and solve them....
Geometry and measurement	Instruction improves the pupils' skills in perceiving the three-dimensional environment and observing elements of plane geometry in it.... The pupils are guided to find and name properties that are used to classify objects and plane figures....	The pupils build, draw, examine, and classify objects and figures.... They learn more about triangles, quadrangles, and circles.... They practice drawing, measuring, and classifying angles.... The pupils are familiarized with the concept of scale, which is applied to enlargements, and reductions....	The pupils expand their understanding of the concepts of point, line segment, straight line, and angle, and familiarize themselves with the concept of the line and ray.... They reinforce their understanding of the concepts of similarity and congruence..... The pupils calculate the circumferences and areas of polygons. Three-dimensional figures are examined....
	The pupils practice measuring, and they are guided to grasp the principle of measurement.	The pupils practice measuring and pay attention to the accuracy of the measurement, estimation of the measurement result, and verifying measurements.	

(continued)

TABLE 4.1 Examples of the Central Content at Different Grade Levels in Comprehensive Schools (continued)

Content Area	Grades 1–2	Grades 3–6	Grades 7–9
Functions			Correlations are depicted both graphically and algebraically.... Pupils get acquainted with the concept of the function....
Data processing, statistics and probability	The pupils begin to develop their ability to collect and store information on interesting topics. The pupils draw and interpret simple tables and bar graphs.	The pupils develop their skill in systematically collecting data on topics that are of interest to them..... They familiarize themselves with probability in everyday life situations by concluding whether an event is impossible, possible, or certain.	The pupils deepen their skills in collecting, structuring, and analyzing data.... They practice defining frequency, relative frequency, and median. They calculate probability.

to utilize programming as part of their studies in different school subjects, with objectives of programming included in mathematics.

Numbers and operations are in the core of primary level mathematics. For example, the curriculum at Grades 1–2 states that students examine properties of numbers besides basic operations and familiarize themselves with the concept of fraction. The curriculum outlines clearly the number range within which the pupils are to learn basic operations, especially during the early years. The basic operational skills and understanding of number concepts, as well as understanding the base ten system, are in the core of the mathematics curriculum. The range of numbers is expanded to real numbers during the lower secondary level starting from Grade 7. An interesting detail is that the division algorithm is no longer included in mathematics at any grade level. Instead, students learn about the relation that division has with other basic operations, and utilize the division algorithm only in the cases that short division can be carried out by dividing each number unit at a time, for example, 248:2.[2]

Algebra comprises its own sub-category within the mathematics curriculum starting only from Grade 3. This differs from the previous national core curriculum that used the concept of algebra from the beginning of compulsory education. However, the same pre-algebraic elements are included in the current curriculum throughout the school years. During Grades 3–6, students get familiar with number sequences and extending them according to a rule. They are introduced to the concept of unknown and they investigate equations. Proportional reasoning, as part of the development of algebraic thinking, cannot be found in the prescription for Grades 3–6, despite the concept of scale mentioned within geometry content (see Table 4.1). Hence, algebra-related reasoning and algebraic thinking are weakly addressed during the early grade levels. In contrast, the content related to algebra and functions for Grades 7–9 mainly follows the traditional mathematics education.

Geometry and measurement is included in the curriculum throughout the school years. Teaching and learning geometry starts with examining the three-dimensional world, from which elements of plane geometry are to be found. Examining objects and plane figures, and analyzing properties of the objects, starts at Grades 1–2. Measuring is practiced especially during the early school years. Concrete models and considering children's perceptions of their environment are seen as the basis for geometry instruction at elementary grades. Only at Grades 7–9 does the curriculum describe geometrical concepts in more detail, showing particular content areas students are to perceive and understand.

In addition to the previous mathematical areas, *data processing*, and *statistics and probability* (at Grades 7–9) comprise a field within school mathematics. It is noteworthy that the curriculum pays attention to pupils' interests and a need to combine instruction with their experiences. For example, data collection and analysis are to be related to topics that are of interest to students. The central mathematical content is described in more detail in Grades 7–9, although the description still uses expressions, such as "pupils examine," "pupils exercise," and "pupils are guided," that actually steer teachers' and students' actions in classrooms.

Learning Environment and Methods

In addition to content, the curriculum also explicitly outlines the nature of the learning environment and even some working methods that are seen as especially appropriate in teaching and learning mathematics. Teaching mathematics is founded on topics and phenomena of interest for pupils throughout the school years. Grades 7–9 mathematics is to provide also problems related to these interests. The aim is to create a learning environment where students study mathematics actively and with the help of various appropriate tools. Utilizing various teaching and learning methods is seen to improve the quality of mathematics education. Students mathematize and solve problems individually and in groups; within teamwork, every member works for her own and for the team's best.

The curriculum emphasizes the essential role of concretizing mathematics: instruction encourages students in using drawings and manipulatives as support for their thinking in Grades 7–9 as well as in earlier grades. Because pedagogically designed play and games are seen to motivate students, these are considered important teaching methods. It is noteworthy that concrete approaches are seen as essential for learning mathematics and thus, materials and manipulatives need to be easily available. Furthermore, the students' role is not only to receive instruction and carry out activities, but also to have an opportunity to influence, for example, the choice of working methods. In addition, digital tools provide meaningful supplements to traditional teaching and learning methods. In Grades 7–9, spreadsheet and dynamic geometry programs are tools to support learning, production, creativity, and evaluation of the work.

Supervision, Differentiation, and Support

Continuous assessment, carried out together with students, is the basis for differentiation of teaching. The curriculum sets high expectations on teachers' capabilities to meet the needs of various learners. Students with learning difficulties are to be offered support, both to catch up with their lack of sufficient knowledge and skills, and to learn new content at the same time. Starting from Grade 3, each student has an opportunity to have instruction in the central content of the earlier grades if the knowledge and skills are inadequate. Systematic *preventive support* to learn new content is a way to support such students. In addition to mathematical knowledge and skills, a possibility of feeling happy about learning and knowing, and opportunities to experience success are particularly mentioned as ways to improve positive attitudes towards mathematics and self-confidence of all students. Appropriate learning materials and manipulatives are an important part of meaningful learning and students also need to have an opportunity to arrive at insights and understanding by themselves. Especially during the later school years, students are seen as active partners in the teaching and learning process. The curriculum discusses the importance of understanding the content from pupils' perspectives, especially at Grades 7–9.

Differentiation also addresses the needs of talented students by offering them opportunities to gain a deeper understanding of the content covered in the particular grade level. Later on, differentiation and special guidance can be organized through using alternative methods and tasks, such as demanding and creative problems, various forms of projects, and problem-based investigations on mathematical topics that are of interest to them.

Assessment and Feedback

The main task of assessment is to support and promote the development of pupils' mathematical thinking in all areas. Pupils' existing skills and possible differences in competences are surveyed at the beginning of the school year. This makes it possible to provide such support that pupils need for completing inadequate previously acquired skills and knowledge. The curriculum underlines the cumulative nature of mathematics that makes it especially important for students to master basic mathematical concepts and skills for later studies. Assessment aims at encouragement and providing feedback to students to utilize their mathematical strengths and further develop their skills in all grade levels. Students become aware of their own learning process; from Grade 3, students are to learn to work persistently and to develop their self-knowledge as learners of mathematics. "Through self-evaluation students learn to set up goals for their own learning and follow their progression in relation to the goals" (FNBE, 2016, p. 405). Students begin to understand, with their teacher's help, what kind of knowledge and skills they need to develop and how. Students learn also about their personal ways of studying and learning mathematics, as well as about their own attitudes toward mathematics, especially at the lower secondary level.

The curriculum enables a teacher to choose various assessment forms to investigate students' levels of mathematical knowledge and skills, for example, through conversations, using manipulatives, drawings, or written products. It states that students have the right to show their knowledge and skills in different ways, both on assessments administered by a teacher or on self-assessment by students. Not only correct solutions but mathematical processes, such as how a student works on tasks, matter. Especially during

Grades 7–9, the character of performance, how students motivate their solutions, how the solutions are structured, and the validity of the solutions are taken into consideration in assessment. In the description of the evaluation for Grades 7–9, the curriculum states that "versatile evaluation and supportive feedback is used to support the development of the pupils' mathematical thinking and self-confidence and to maintain and strengthen their learning motivation" (FNBE, 2016, p. 405). Students are to be informed regularly about their progress and performance in relation to the learning objectives in mathematics. Evaluation aims at guiding each student to develop her knowledge and understanding.

The curriculum outlines the essence of assessment for different grade levels regarding mathematical content. During Grades 1–2, assessment focuses especially on progress in understanding number concepts and number sequences, the base ten system and basic operations, classifying geometric bodies and figures, as well as using mathematics in problem solving. The focus of assessment is outlined in more detail during the later school years following the content areas described in the section on mathematical content.

Assessment also concerns group work. The aim is to help students understand the significance of each team member's work and development. The working process and results of the whole group, as well as the contribution of each group member, are equally important and evaluated in Grades 3–9. Students learn to assess the process of working together and its results.

Upper Secondary School Mathematics

The Finnish curriculum of upper secondary mathematics emphasizes the value and role of mathematics in different fields of society, such as in science, technology, the economy, and everyday lives of individual citizens. Thus, skills of problem solving and modeling are seen as important goals for studying mathematics in addition to conceptual and procedural competences. Mathematics is seen as a tool for constructing and gaining special knowledge in society; and especially for the advanced level, understanding of the structure and nature of pure mathematics as a field of science is given emphasis. The current digitalization in education is realized in the upper secondary school mathematics curriculum so that, for every mathematics course, there is an objective that involves the use of technical tools for investigating the important concepts and solving problems in that particular mathematical area. The technical tools include programs of dynamic geometry, symbolic calculation, spreadsheets, statistics, and word processing, all of which will be available for students in the final matriculation examinations beginning in spring 2019. It will be seen in a few years how important the role of technology plays out in the matriculation examination of mathematics in the future.

According to the national core curriculum, the first course in mathematics is a common compulsory course for both mathematical levels (basic and advanced) in order to attract more students than before to choose the advanced level mathematics (see Table 4.2). The content of the course is real numbers, operations, percentages, functions, number sequences, as well as powers and logarithms. When compared to the previous syllabus, the content of the advanced level courses is almost the same. Only some minor revisions have been done to the order and content of the courses in the recent curriculum reform, for example, the inclusion of functions of two variables and their partial derivatives in the last optional course on analysis. A more extensive reorganization has taken place in the content of the basic level mathematics courses. Analysis is no longer included in the compulsory studies; instead, a national optional course is devoted to the topic. The national optional course on vectors and trigonometry has been removed from the syllabus. A course on mathematical economics is now included in the compulsory part for all students, instead of just being an optional national course. Also, the status of statistics and probability is strengthened by adding a compulsory course and a national optional course on that particular topic.

Support of the Reform

Although FNBE provides some material for teachers as well as parents to support the curricular reform, there is no tradition of systematically supporting teachers' professional development or curriculum reforms on a large scale. Instead, various regional authorities, universities, teachers' unions, municipalities, different associations, companies, and groups of individuals have received grants and organized both ad hoc and long-term projects for inservice teacher training.

LUMA Centre Finland (LU abbreviated from *luonnontieteet*, the Finnish word for natural sciences, and MA *mathematics* www.luma.fi/centre) was established in 2013 as an umbrella organization for regional LUMA Centres in Finnish Universities. Today, 13 universities collaborate to inspire and motivate 3–19 year old students and their teachers to study topics of science, mathematics, technology, and information technology. Preservice and inservice teacher training are important ways for LUMA Centres to work with teachers. These Centres have developed different profiles to promote their common mission. LUMA Centre Finland has launched a national wide development project with various sub-projects for renewing the teaching of mathematics and science in the comprehensive

TABLE 4.2 Examples of the Core Objectives and Content of Mathematics in the General Upper Secondary School

Course	Objective of the Course	Core Content
Common Compulsory Course		
Numbers and number sequences	• Student reflects on the significance of mathematics from the perspective of the individual and society. • Student deepens his or her understanding of the concept of function. • Student is able to determine number sequences when provided the premise and method of forming the terms that follow.	• Real numbers, basic arithmetic operations, and percentage calculations • Functions, drawing, and interpreting graphs • Number sequences • Arithmetic progression and sums • Solving equations in the form of $a^x = b, x \in \mathbb{N}$ • Geometric progression and sums
Mathematics: Basic Syllabus		
Expressions and equations	• Student gains practice in using mathematics to solve everyday problems and learns to trust his or her own mathematical abilities. • Student advances his or her equation-solving skills and learns to solve quadratic equations. • Student is able to use technological tools to examine polynomial functions and to solve application problems related to polynomial equations and polynomial functions.	• Linear dependency and proportionality between quantities • Converting verbal problems into equations • Interpreting and assessing solutions • Quadratic polynomial functions and solving quadratic equations
Geometry	• Student gains practice in making observations and drawing conclusions about the geometrical properties of figures and bodies. • Student knows how to solve practical problems using geometry. • Student is able to use technological tools to examine figures and objects and to solve application problems related to geometry.	• Similarity of figures • Pythagoras' theorem and the converse of the Pythagorean theorem • Determining area and volume of figures and bodies

(continued)

Mathematics Curriculum: The Case of Finland ■ 91

TABLE 4.2 Examples of the Core Objectives and Content of Mathematics in the General Upper Secondary School (continued)

Course	Objective of the Course	Core Content
Mathematical models	• Student perceives regularities in and dependencies between real-world phenomena and describes these using mathematical models. • Student learns to assess the quality and practicability of different methods. • Student familiarizes himself or herself with making predictions based on models.	• Applying linear and exponential models • Solving exponent equations • Solving exponent equations using logarithms • Number sequences as mathematical models
Statistics and probability	• Student gains practice in processing and interpreting statistical material. • Student evaluates different regression models, for example with spreadsheet software, and makes predictions with the help of models. • Student familiarizes himself or herself with the basics of probability theory.	• Concepts of regression and correlation • Observation and deviating observation • Making predictions • Combinatorics • The concept of probability
Commercial mathematics	• Student advances his or her skills in percentage calculation. • Student understands the concepts used in the context of economics. • Student strengthens his or her mathematical foundation for studies in entrepreneurship and economics. • Student applies statistical methods to the processing of data.	• Index, cost, money transaction, loan, tax, and other calculations • Mathematical models applicable to economic situations, using number sequences and sums
Mathematical analysis (optional specialization course)	• Student examines the rate of change of a function using graphical and numerical methods. • Student understands the concept of derivative as a measure of the rate of change. • Student is able to determine the greatest and smallest value of a polynomial function in connection with applications.	• Graphical and numerical methods • Derivatives of polynomial functions • Examining the sign and behavior of a polynomial function • Determining the maximum and minimum of a polynomial function

(continued)

TABLE 4.2 Examples of the Core Objectives and Content of Mathematics in the General Upper Secondary School (continued)

Course	Objective of the Course	Core Content
Statistics and probability II (optional specialization course)	• Student strengthens and diversifies his or her skills when processing and interpreting statistics. • Student is able to define statistical parameters and probabilities with the help of continuous distributions, utilizing technical tools.	• Normal distribution and concepts of standardization of distributions • Binomial test • Binomial distribution • The concept of the confidence interval
Mathematics: Advanced Syllabus		
Polynomial functions and equations	• Student gains practice in using polynomial functions. • Student is able to solve higher order polynomial equations and examine the number of possible solutions. • Student is able to solve simple polynomial inequalities. • Student is able to use technical tools in examining polynomial functions and in solving application problems related to polynomial equations, polynomial inequalities, and polynomial functions.	• Products of polynomials and the binomial theorem in the form of $(a+b)^n$, $n \leq 3$, $n \in \mathbb{N}$ • Quadratic equations, its formula, and examining its number of roots • Factorization of quadratic polynomials • Polynomial functions • Polynomial equations • Solving polynomial inequalities
Geometry	• Student gains practice in perceiving and describing information about the space and shape in both two and three dimensions. • Student gains practice in formulating, justifying, and using statements dealing with geometrical information. • Student is able to use technical tools to examine figures and objects and in solving application problems related to geometry.	• Similarity of figures and objects • Sine and cosine theorems • Geometry of a circle, its parts, and straight lines related to it • Calculating lengths, angles, areas, and volumes related to figures and objects

(continued)

TABLE 4.2 Examples of the Core Objectives and Content of Mathematics in the General Upper Secondary School (continued)

Course	Objective of the Course	Core Content
Vectors	• Student understands the concept of the vector and familiarizes himself or herself with the basics of vector calculus. • Student understands the principle of solving systems of equations. • Student learns to examine points, distances, and angles in two- and three-dimensional coordinate systems by means of vectors.	• Basic properties of vectors • Addition and subtraction of vectors and scalar multiplication of vectors • The scalar product of vectors in the coordinate system • Solving systems of equations • Straight lines and planes in space
Analytical geometry	• Students understand how analytical geometry links geometric and algebraic concepts. • Student understands the concept of the equation of a set of points and learns to examine points, straight lines, circles, and parabolas using equations. • Student deepens his or her understanding of the concept of absolute value and learns to solve simple absolute value equations and corresponding inequalities of the form $|f(x)| = a$ or $|f(x)| = |g(x)|$.	• Equation of sets of points • Equation of straight lines, circles, and parabolas • Solving absolute value equations and inequalities • Distance from a point to a straight line
Derivative	• Student knows how to define the zeros of a rational function and solve simple rational inequalities. • Student is able to determine derivatives of simple functions. • Student knows how to determine the greatest and smallest value of a rational function. • Student is able to use technical tools in examining limits, continuity, and derivatives, solving rational equations and inequalities....	• Rational equations and inequalities • Limits, continuity, and derivatives of functions • Differentiation of polynomial functions and of the products and quotients of functions • Examining the behavior of a polynomial function and determining its extremum

(continued)

TABLE 4.2 Examples of the Core Objectives and Content of Mathematics in the General Upper Secondary School (continued)

Course	Objective of the Course	Core Content
Trigonometric functions	• Student examines trigonometric functions using the symmetries of the unit circle. • Student is able to differentiate composite functions. • Student is able to utilize trigonometric functions in modelling sequential phenomena. • Student is able to use technical tools in examining trigonometric functions, solving trigonometric equations, and determining the derivatives of trigonometric functions in problem-solving assignments.	• Directed angle and radian • Trigonometric functions, including their symmetric and periodic properties • Solving trigonometric equations • Derivatives of composite functions • Derivatives of trigonometric functions
Radical and logarithmic functions	• Student is familiar with the properties of radical, exponentiation, and logarithmic functions and knows how to solve equations related to these. • Student is able to examine radical, exponential, and logarithmic functions by means of the derivative. • Student is able to utilize the exponential function in modelling different phenomena of increase and decrease.	• Rules of calculating exponents • Radical functions and equations • Exponential functions and equations • Logarithmic functions and equations • Derivatives of radical, exponential, and logarithmic functions
Integral calculus	• Student understands the concept of the definite integral and its connections to the area. • Student learns to determine area and volume by means of definite integrals. • Student familiarizes him- or herself with the applications of integral calculus.	• Integral function • Integral functions of elementary functions • The definite integral • Calculating area and volume
Probability and statistics	• Student familiarizes himself or herself with combinatorial methods. • Student understands the concept of discrete probability distribution and learns to define and apply the expected value of the distribution. • Student familiarizes himself or herself with the concept of continuous probability distribution and learns to apply the normal distribution.	• Discrete and continuous statistical distributions • Distribution parameters • Combinatorics • Rules of calculating probabilities • Normal distribution

(continued)

TABLE 4.2 Examples of the Core Objectives and Content of Mathematics in the General Upper Secondary School (continued)

Course	Objective of the Course	Core Content
Number theory and mathematical proofs (optional specialization course)	• Student familiarizes himself or herself with the basis of logic and the principles of mathematical proof, and practices providing proof. • Student is able to examine the divisibility of integers by means of division equations and the congruence of integers. • Student deepens his or her understanding of number sequences and their sums.	• Connectives and truth-values • Providing geometric proof • Induction proof • Euclidean algorithm • The Fundamental Theorem of Arithmetic
Algorithms in mathematics (optional specialization course)	• Student enhances his or her algorithmic thinking. • Student is able to examine and explain how algorithms work. • Student is able to determine rates of change and areas numerically.	• Iteration and Newton-Raphson's method • Newton-Cotes Formula: rectangle rule, trapezoid rule, and Simpson's rule
Advanced differential and integral calculus (optional specialization course)	• Student deepens his or her knowledge of the theoretical foundations of differential and integral calculus. • Student examines the inverse functions of strictly monotone functions. • Student is able to examine the limits of number sequences, series, and their sums.	• Examining the continuity and differentiability of functions • General properties of continuous and differentiable functions • The function of two variables and partial derivatives • The limits of functions and number sequences in infinity

schools (6 to 16 year old students) in line with the new curriculum. The results of the most successful sub-projects will be spread to schools in Finland during a three-year period from 2017–2019.

One example of the need for inservice teacher education is *programming* in the new mathematics curriculum from Grades 1 to 9. Similar to general inservice teacher training, courses for teaching programming are organized by different agents. The Innokas project is a nationwide initiative to promote technology-based innovation education, with a special focus on robotics. Further, universities and groups of teachers have organized online courses for teachers to learn different types of programming. One *grassroots initiative* is an open online course in programming for teachers called Koodiaapinen (Code Alphabet; http://koodiaapinen.fi/en/), driven by teachers and educational researchers, who mostly voluntarily contribute to developing an open web-based inservice course (MOOC). For the first term in autumn 2015, 36% of 2,727 persons who registered for the course finished it. The organizers received a great deal of feedback from the participants to improve the course for the next term. Even volunteers from information technology and games companies have given their support to schools and teachers. Unfortunately, teaching programming has been much technology-driven. It means that the focus has been on considering how to apply a particular tool in mathematics classrooms without developing a broader view of teaching and learning programming. A holistic picture of the learning path of children is still waiting to be sketched out, as municipalities and schools write their own curriculums.

The prescriptions of the new mathematics curriculum pose certain demands on the schools concerning locally available concrete materials and technical support as well as on inservice teacher education. Already the previous curriculum (2004) brought about the use of concrete materials when learning and exercising mathematics. This led to development by municipalities (sometimes in co-operation with universities) of specific inservice centers for teachers and students, called Math Wonderlands, which offered professional development courses concerning the use of concrete materials in teaching. Yet, this support is not systematic in all parts of the country.

The commercial textbook companies have for several decades produced concrete materials that are attached in students' textbooks (Grades 1–6), a practice that has obviously enhanced the equality of mathematics education. They have also offered a variety of ideas for activities and games to be used in mathematics learning, at least for Grades 1–6. Although the companies are currently producing more and more digital teaching and learning materials, many teachers in the field still need inservice education, for example, in how to productively use different programs in their teaching. How school authorities will meet the demands of the use of digital tools in the current economically strained situation in Finland is not clear.

The new curriculum also poses high expectations for continuous assessment and differentiation of teaching, both by offering support for struggling students and challenges to students who are able to learn more than the average group. How this should be accomplished is not exemplified. However, formative assessment, differentiation, and supporting struggling students are important areas of study that have been addressed in Finnish teacher education for quite some time. There is also a long tradition among commercial textbook publishers to attempt to meet this challenge. They often organize materials in a way that allows teachers to flexibly use a textbook for different learners while simultaneously keeping students studying the same mathematical topic.

DISCUSSION

Concerning the aspect of decentralization versus national steering, the new curriculum document for comprehensive schools seems to be more steering with respect to the teaching methods used in mathematics classrooms than the previous curricula. There are no criteria established for good performance for the early grades as in the previous curriculum. Hence, the steering is slightly weaker with respect to the prescribed student progression in mathematics.

Mathematical content and goals are also embedded in the prescriptions about instructional approaches and they are to be interpreted and enacted in the frame of the transversal (generic) competence areas. An interesting feature in the Finnish curriculum is that there is an aim to enhance students' interest towards learning and school by working in an interdisciplinary manner with various projects that interest students. However, nothing about the history of mathematics is included in the mathematics curriculum at any level, something that could help students see human beings behind mathematics and maybe awaken the interest of students who are more oriented towards humanities and the social sciences. Still, it depends on a school and teachers how essential a role mathematics will have in these multidisciplinary activities.

The content and goals for at least Grades 1–6 are mathematically less demanding compared to the earlier 2004 curriculum. Algebra is not mentioned in Grades 1–2 and inequalities have been omitted in the content of Grades 3–6. Proportional reasoning, which is strongly connected to the development of algebraic thinking, is not mentioned in the new curriculum for Grades 1–6. Division and multiplication of rational numbers have been postponed to Grades 7–9. Geometry includes fewer concepts for students to learn than previously, and teaching and learning geometry starts with notions of students' three-dimensional environments. The content and goals in geometry are very similar to those traditionally connected to preschool classes, even though there are no obligatory aims to steer learning outcomes during the preschool year in Finland. Only within the measurement domain some concepts are included that may be new for children ages 7–8 (Grades 1–2). All these changes are interesting against the background that preschool is now obligatory for all children.

The weakening of the algebra prescriptions for Grades 1–6 is interesting for several reasons. A great body of didactical research shows that it is beneficial for young students to work with big ideas in algebra before more formal treatment during adolescence, and there are promising approaches to early algebra described in research reports (e.g., Blanton et al., 2015). Moreover, besides geometry, algebra and competencies needed for the development of algebraic thinking, such as identifying patterns, dealing with fractions and generalization, percentages, and handling of algebraic expressions, have been identified as especially problematic for Finnish students in recent international and national evaluations (Kupari, 2013).

The new curriculum puts considerable emphasis on teaching methods (for example digitalization), continuous evaluation, differentiation, support of struggling students, and motivational aspects of learning. The previous curriculum also addressed joy and students' interest and set demands for the use of information and communications technology (ICT). Now these aspects dominate the mathematics curriculum. The new curriculum steers teachers to guide students in their individual work and group work rather than explaining mathematics for them. The words *supervise* and *guide* are used 48 times in the mathematics curriculum, while the word *teach* is used only once, and *introduce* a couple of times. Another quite popular term is *support*, used 22 times, and the word *encourage* used especially in the description of the curriculum for Grades 7–9.

Overall, the process orientation (Eisner, 1985) is the most visible orientation in the reform document. Students are to *examine* and *discover*, individually and in groups. They are also to *learn how to learn*. Students need to have an opportunity to learn about and choose their working manner, and good mathematics instruction is seen to be based on phenomena of students' interest. Hence, the orientation is to some extent also that of personal relevance (Eisner, 1985). Although the teachers' responsibility for students' interest is addressed, basing teaching solely on students' interest is not expected. The orientation of social adaptation (Eisner, 1985) is clearly addressed in the new curriculum, for example by including programming in mathematics. The demands of digitalization of teaching and learning can also be seen as a manifestation of social adaptation. But these features can be connected to the orientation of social reconstruction (Eisner, 1985), as they aim to provide students with competences needed in a changing world.

The view of mathematics is quite balanced as to the demands of conceptual and procedural knowledge and skills. Attributes such as *cumulative, accurate, logical,* and *creative* are addressed in the document, and understanding of concepts, and structures between the concepts, as well as procedures are important goals. The word *understand/understanding* is central in the document, as it occurs equally spread in the prescriptions of the different grade levels, in all 54 times. Also, knowing and mastering certain skills are addressed.

Problem solving and students' investigations are ascribed a prominent place in the document and are in line with recent trends in mathematics education (cf. e.g., Stein & Smith, 2011), in which students present their solutions and discuss different ways of solving problems. In contrast to the international trend in mathematics education (e.g., Stylianou, Blanton, & Knuth, 2009), proof and proving seem to be weaker in emphasis in the new comprehensive school curriculum than in the previous one, and there is no visible change in the upper secondary curriculum (Hemmi et al., 2013). Some general proof-related competences, such as draw conclusions (Grades 1–2), draw motivated conclusions (Grades 3–6), and motivate and draw conclusions (Grades 7–9), are mentioned in the description of *mathematical thinking*. In addition, it is stated that students learn to critically review how reasonable the solution is (Grades 3–6 and 7–9), and engage in activities demanding logical thinking, such as finding a rule and dependence relations and presenting them in an exact manner (Grades 7–9). Mathematics education aims also at getting insights to basics of proving and exercise to judge the truth-value of statements during the lower secondary school.

CONCLUSION

The value of equal opportunity and a vision of lifelong learning throughout the system have been consistent objectives for 40 years in Finland. High-quality teachers with strong academic, research-based teacher education, local responsibility for educational quality, and a support system for different learners have produced consistently high student learning outcomes in all Programme for International Student Assessment (PISA) measurements (Niemi et al., 2014). Adult competences have also been at the top of the curriculum. Now we face several challenges, among them the rapid weakening of students' knowledge and skills in mathematics, as well as their negative attitudes. There are also changes in Finnish society, such as immigration, an aging population, and new kinds of jobs that put new demands on the curriculum. The number of immigrant students is growing in Finland and the differences between their results compared to students speaking native languages are huge compared to many other countries (OECD, 2013).

The new national core curriculum aims to crystallize the vision of education for the future and the expertise needed in Finnish society, thus it is based on 21st century skills and competences. The mathematics curriculum for compulsory school conveys a strong willingness to change the trend concerning students' attitude and interest as well as their self-efficacy concerning mathematics. This may be accomplished, for example, by including more games and activities during Grades 7–9. Another means for fulfillment of this goal is cooperative and self-regulated learning. Even if students take responsibility for their own learning, it is the teacher's duty to motivate and model enthusiasm, and spur students to maintain their interest and willingness to learn.

The current reform with respect to mathematics is not based on profound analyses about why Finnish students' results on international assessments have been outstanding for some time and why we now witness a down-going trend. There have only been separate attempts to explain the earlier success (Andrews, Ryve, Hemmi, & Sayers, 2014). It will be interesting to follow how the school authorities, textbooks, and teacher guides, as well as teachers, interpret the new curriculum and what kind of impact it will have on students' interest towards, and their learning outcomes in mathematics.

NOTES

1. Functions are in the focus only during Grades 7–9 with only a short paragraph in the curriculum related to this content area. Therefore, we do not discuss this content area in depth in this chapter.
2. 248:2 stands for "248 divided by 2" and is a common notation for division in Finnish school mathematics.

REFERENCES

Andrews, P., Ryve, A., Hemmi, K., & Sayers, J. (2014). PISA, TIMSS and Finnish mathematics teaching: An enigma in search of an explanation. *Educational Studies in Mathematics, 87*(1), 7–26. doi:10.1007/s10649-014-9545-3

Blanton, M., Stephens, A., Knuth, E., Murphy Gardiner, A., Isler, I., & Kim, J.-S. (2015). The development of children's algebraic thinking: The impact of a comprehensive early algebra intervention in third grade. *Journal for Research in Mathematics Education, 46*(1), 39–87.

Eisner, E. W. (1985). Five basic orientations to the curriculum. In E. W. Eisner (Ed.), *The educational imagination* (pp. 61–86). New York, NY: Macmillan.

Finnish National Board of Education (FNBE). (2004). National core curriculum for basic education 2004. Retrieved in December 2016 from http://www.oph.fi/english/curricula_and_qualifications/basic_education/curricula_2004.

Finnish National Board of Education (FNBE). (2016). National core curriculum for basic education 2014. Publications 2016:5.

Hemmi, K., Lepik, M., & Viholainen, A. (2013). Analysing proof-related competences in Estonian, Finnish and Swedish mathematics curricula—towards a framework of developmental proof. *Journal of Curriculum Studies, 45*(3), 354–378. doi:10.1080/00220272.2012.754055

Joutsenlahti, J., & Vainionpää, J. (2010). Oppimateriaali matematiikan opetuksessa ja osaamisessa [Learning materials in the teaching and learning of mathematics]. In E. K. Niemi & J. Metsämuuronen (Eds.), *Miten matematiikan taidot kehittyvät? Matematiikan oppimistulokset peruskoulun viidennen vuosiluokan jälkeen vuonna 2008* [How do pupils' mathematical skills develop? The learning outcomes in the end of the fifth grade in compulsory school] (pp. 137–146). Helsinki, Finland: Opetushallitus (The Finnish National Board of Education).

Ketamo, H. (2014). Learning by teaching: A game-based approach. In H. Niemi, J. Multisilta, L. Lipponen, & M. Vivitsou (Eds.). *Finnish innovations and technologies in schools. A guide towards new ecosystems of learning* (pp. 77–85). Rotterdam, the Netherlands: Sense.

Kumpulainen, T. (Ed.). *Koulutuksen tilastollinen vuosikirja 2014. Koulutuksen seurantaraportit 2014:10 Opetushallitus.* [Statistical Yearbook of Education 2014. Education Report Series 2014:10 FNBE]. Retrieved August 2016 from http://www.oph.fi/download/163331_koulutuksen_tilastollinen_vuosikirja_2014.pdf

Kulm, G., & Li, Y. (2009). Curriculum research to improve teaching and learning: National and cross-national studies. *ZDM—The International Journal on Mathematics Education, 41*(6), 709–715.

Kupari, P. (1993). Matematiken i den finska grundskolan. Attityder och kunskaper. [Mathematics in the Finnish comprehensive school—attitudes and achievements]. *Nordisk Matematikdidaktik, 1*(2), 30–58.

Kupari, P. (2013). *Perusopetuksen matematiikan oppimistulokset ja niiden kehitys* [Learning outcomes and their evolution in mathematics during basic education]. Jyväskylä, Finland: Finnish Institute for Educational Research. Retrieved from https://ktl.jyu.fi/pirls-timss/timss/maol_13.

Kupari, P., Sulkunen, S., Vettenranta, J., & Nissinen, K. (2012). *Enemmän iloa oppimiseen. Neljännen luokan oppilaiden lukutaito sekä matematiikan ja luonnontieteiden osaaminen. Kansainväliset PIRLS- ja TIMSS-tutkimukset Suomessa* [More pleasure in learning. Fourth-graders' knowledge and skills in literacy, mathematics, and science. International PIRLS and TIMSS investigations in Finland]. Jyväskylä, Finland: Finnish Institute for Educational Research.

Lavonen, J. & Krzywacki, H. (2010). Implementation of Finnish education policy through national core curriculum: Science as an example. In J. Kirylo & A. Nauman (Eds.), *Curriculum development: Perspectives from around the world* (pp. 118–131). Olney, MD: Publications for the Association of Childhood Education International ACEI.

Niemi, E. K., & Metsämuuronen, J. (Eds.). (2010). *Miten matematiikan taidot kehittyvät? Matematiikan oppimistulokset peruskoulun viidennen vuosiluokan jälkeen vuonna 2008* [How do pupils' mathematical skills develop? The learning outcomes in the end of the fifth grade in compulsory school]. Helsinki, Finland: Opetushallitus (The Finnish National Board of Education).

Niemi, H., & Multisilta, L. (2014). Global is becoming everywhere: Global sharing pedagogy. In H. Niemi, J. Multisilta, L. Lipponen, & M. Vivitsou (Eds.), *Finnish innovations and technologies in schools: A guide towards new ecosystems of learning* (pp. 35–48). Rotterdam, the Netherlands: Sense.

Niemi, H., Multisilta, J., Lipponen, L., & Vivitsou, M. (Eds.). (2014). *Finnish innovations and technologies in schools. A guide towards new ecosystems of learning.* Rotterdam, the Netherlands: Sense.

Organization for Economic Cooperation and Development (OECD). (2013). *PISA 2012 results in focus. What 15-year-olds know and what they can do with what they know.* Available on: www.oecd.org/pisa

Pehkonen, L. (2004). The magic circle of the textbook—An option or an obstacle for teacher change. In M. J. Hoines, & A. B. Fuglestad (Eds.), *Proceedings of the Annual Meeting of the International Group for the Psychology of Mathematics Education (PME)* (28th, Bergen, Norway, July 14–18, 2004) (pp. 513–520). ERIC: ED489178

Prytz, J., & Karlberg, M. (2016). Nordic school mathematics revisited—On the introduction and functionality of New Math. *NOMAD, 21*(1).

Rokka, P. (2011). *Peruskoulun ja perusopetuksen vuosien 1985, 1994 ja 2004 opetussuunnitelmien perusteet poliittisen opetussuunnitelman teksteinä* [The foundations of the curricula of compulsory school and basic education in 1985, 1994 and 2004 as political texts] (Doctoral thesis). University of Tampere.

Routti, J., & Ylä-Anttila, P. (2006). *Finland as knowledge economy. Elements of success and lessons learned.* Washington DC: World Bank.

Sahlberg, P. (2011). The fourth way of Finland. *Journal of Educational Change, 12*(2), 173–185.

Simola, H. (2002). The Finnish miracle of PISA: Historical and sociological remarks on teaching and teacher education. *Comparative Education, 41*(4), 455–470.

Simola, H., Rinne, R., Varjo, J., Pitkänen, H., & Kauko, J. (2009). Quality assurance and evaluation (QAE) in Finnish compulsory schooling: A national model or just unintended effects of radical decentralisation? *Journal of Education Policy, 24*(2), 163–178.

Stylianou, D., Blanton, M., & Knuth, E. (2009). *Teaching and learning proof across the grades: A K–16 perspective.* New York, NY: Routledge.

Stein, M. K., & Smith, M. (2011). *5 practices for orchestrating productive mathematics discussions.* Reston, VA: National Council of Teachers of Mathematics.

Törner, G., Potari, D., & Zachariades, T. (2014). Calculus in European classrooms: Curriculum and teaching in different educational and cultural contexts. *ZDM—The International Journal on Mathematics Education, 46*(4), 549–560.

Tuohilampi, L., Hannula, M. S., Laine, A., & Metsämuuronen, J. (2014). Examining mathematics-related affect and its development during comprehensive school years in Finland. In C. Nicol, S. Oesterle, P. Liljedahl, & D. Allan, (Eds.), *Proceedings of the Joint Meeting of PME 38 and PME-NA 36, Vol. 3* (pp. 281–288). Vancouver, Canada: PME.

Vahtivuori-Hänninen, S., Halinen, I., Niemi, H., Lavonen, J., & Lipponen, L. A. (2014). New Finnish national core curriculum for basic education and technology as an integrated tool for learning. In H. Niemi, J. Multisilta, L. Lipponen, & M. Vivitsou (Eds.), *Finnish innovations and technologies in schools. A guide towards new ecosystems of learning* (pp. 21–34). Rotterdam, the Netherlands: Sense.

CHAPTER 5

CURRICULUM IN CANADA

A Fractal Interpretation Using the Case of Alberta

Elaine Simmt
University of Alberta

Mathematics curriculum in Canada is not singular. There is not a national curriculum. However, curriculum is an important part of a national discourse in mathematics education, with collaboration and cooperation among many jurisdictions. There are many influences on curriculum: political and social, personal and professional, local and international.[1] Although curriculum is produced by government policy makers and experts, Canadian mathematics educators and educational researchers influence that development and implementation with their research and their work as teacher educators. But in the end, the mandated curriculum is the responsibility of the government, implementation is the responsibility of school districts and teachers, and living the curriculum is the responsibility of children and youth. With this in mind, I present an interpretation of curriculum in Canada by examining the curriculum from one region, the province of Alberta. In this work, I interpret curriculum as (all at once) historical, cultural, personal, political and politicized, and geographical. It is what I call a fractal interpretation. As with other fractals, when one zooms in on a particular feature, other features come into view. In this way, there is an aspect of self-similarity in my analysis. For each theme, I find the other themes within it (see Figure 5.1).

CONTEXT

There are two features of Canadian curriculum that must be noted before further elaboration, especially for international readers. The first is with the technical use of the word *curriculum*. It is not common for Canadians to refer to a textbook series as *the* curriculum; rather, we speak of textbooks as curriculum resources or materials. For most educators in Canada, the word curriculum is used to denote a program of studies that specifies a set of outcomes and directives mandated by ministries of education for the purpose of contributing to the goals of education.

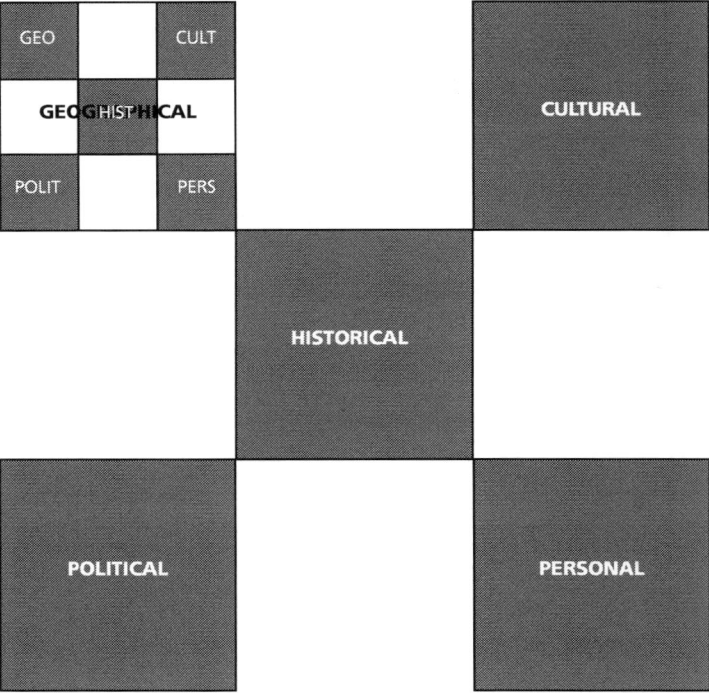

Figure 5.1 Fractal view of curriculum.

The second feature is related to the site of responsibility for curriculum. Through the *Canadian Constitution Act of 1867*, provinces were given legislative powers over education (Council of Ministers of Education, Canada [CMEC], 2008). Hence, education in Canada is a regional responsibility. There is no national curriculum, no national examination, no national teacher education program, and no national certification. Therefore, it is not possible to write about *the* Canadian curriculum. To write about it, one would have to write about 13 different curricula. Rather than doing that, I focus specifically on a single jurisdiction under the premise that mathematics curricula from across the country, their challenges of development and implementation, and their implications for the education of children and youth are similar throughout Canada. In particular, I use the curriculum and curriculum discourse from Alberta as an example of Canadian curriculum.

Alberta is in Western Canada. It is the lead province for mathematics curriculum development under the Western and Northern Canada Protocol for Basic Education (WNCP). A *Common Curriculum Framework (CCF) for K–12 Mathematics* was developed through a collaborative multi-jurisdictional process under the Western Canada Protocol (WCP, 1996) and WNCP (2006, 2008).[2] Four provinces (British Columbia, Alberta, Saskatchewan, Manitoba) and three territories (Yukon, Northwest Territories, Nunavut) participated in the process. Today, the CCF is the basis (fully or in part) for mathematics programs in ten regional jurisdictions across Canada (the WNCP partners, and the eastern provinces of New Brunswick, Nova Scotia, and Newfoundland), all acknowledging the adaptation of the CCF for their mathematics programs. The curricula of Ontario, Quebec, and Prince Edward Island (the remaining provinces of Canada) are not reflected in the discussions in this paper. None of those jurisdictions use the CCF. Rather, each continues to develop and maintain autonomous programs of study for their education system.

FRACTAL VIEW OF CURRICULUM

This paper is a reflection on curriculum in Canada constructed from my experiences as a mathematics teacher, teacher educator, and educational researcher from Alberta, and is informed by the work of my Canadian colleagues. I discuss curriculum as a multidimensional phenomenon that involves different experiences and expectations of multiple stakeholder groups (and the individuals within them) and the governmental responses to those stakeholders as they are manifested in curriculum transformation. I draw inspiration from Canadian curriculum reconceptualists in mathematics education (e.g., Davis, 1996; Jardine, Clifford, & Friesen, 2008), who teach that, when we want to understand curriculum, we must explore it as a complex lived phenomenon, and in doing so we are called to consider its historical, cultural, personal, political, and geographical dimensions. Through these dimensions, I explore different aspects of curriculum: conceptual understandings and interpretations, processes of development, and dynamics of implementation.

There is a fractal nature to my analysis that begins with a historical account of curriculum reform in Western Canada from the 1940s to the present. I am sensitized by my observation that, when I analyse any one dimension of curriculum, the other dimensions are found within it and extend from it. In other words, as I focus on the curriculum from a historical perspective, the cultural, personal, political, and geographical dimensions of curriculum emerge as co-implicated features. There are moments when I use examples within one dimension and then return to them when discussing another dimension of curriculum.

HISTORICAL FEATURES OF CURRICULUM

Since 1904 when education in Alberta was enacted through legislation as a provincial responsibility, mathematics education has undergone continuous transformation. Clearly, like all institutions, education is not a singularity within a society but is impacted by many influences and reflects the societal goals and values.

Cycles of Review, Revise, and Reform

Documentation from the last 100 years reveals cycles of curriculum revision triggered by questions of the relevance and effectiveness of the mathematics curriculum, which were followed with reviews and revisions of the program of studies, leading to the implementation of new curriculum. Over time, the frequency with which these cycles have occurred has increased, with a greater number of years between cycles in the early 20th century to fewer years between them in recent decades.[3] The cycles of reform have resulted in modest changes in some periods and more dramatic ones in others. In Alberta, the 1940s, 1960s, 1990s, and 2000s were periods of dramatic reforms with the addition of courses to make mathematics education more inclusive, deletion of courses that did not live up to expectations, and alteration of high school graduation requirements. Even since the last significant change in 2007, the curriculum has been undergoing continuous minor revisions (which some people feel are significant). I return to this issue in the section on political dimensions of curriculum.

In the earliest years of mathematics education in Alberta, the curriculum consisted largely of arithmetic and practical mathematics for primary school (topics needed for day to day life), and algebra, geometry, applied arithmetic, and trigonometry for secondary school (topics needed for university education or commercial pursuits). In the early 20th century, many learners did not complete secondary school. However, the 1940s saw the addition of a general mathematics course for high school students who were not studying for matriculation. Mid-century in Alberta (1961), graduation requirements were changed from completion of Grade 8 Mathematics to Grade 11 Mathematics (a requirement that remains in place today in Alberta). In the 1990s, one of the most significant program changes to Alberta mathematics curriculum was the introduction of a stream in applied mathematics. This course was to "emphasize applications of mathematics rather than precise mathematical theory" and the "approaches used are primarily numerical and geometrical" (WCP, 1996, p. 19). As discussed later in the curriculum as political and politicized, the applied mathematics stream had a short life in Alberta.

Aims and Goals of Mathematics Education

Since the earliest programs of study in Alberta, the aims of education have focused on the child. In the 1940s the influence of the progressive school movement is evident:

> ... [to] facilitate the child's progressive orientation in the life of which he is a part; to provide an environment that sustains growth and development; to promote social adjustment; to develop desirable attitudes, ideals and appreciations; to develop necessary skills and to impart information; and to promote health, both physical and mental. (Department of Education,[4] 1940, p. 3)

Teachers of mathematics were instructed, "Early number teaching must be informal...it must be determined by the maturity of the child, and be guided by his interest" (p. 255).

In the 1960s Alberta, like other jurisdictions, was sensitive to the needs of society post Sputnik. Mathematics education, in particular, was expected to contribute to the modern education of children by

- supporting their "powers of clear understanding and logical thinking";
- encouraging their habits of "critical thinking and intellectual honesty";
- contributing to their appreciation of the "contribution of mathematics to the progress of civilization";
- developing and maintaining their "numerical computation skills and processes"; and

- leaving them with "a sense of satisfaction from accomplishment and a sense of personal responsibility for accuracy, neatness and precision" (Department of Education, 1963a, p. 4).

In sum, mathematics was expected to serve as a "guide to intelligent economic living and social adjustment" (p. 4).

As was seen in previous decades, the triad of the individual, society at large, and the discipline of mathematics underpinned the aims of mathematics education in the 1980s. However, it was this decade that saw the influence of the psychology of mathematics education (and early constructivist research) in the curriculum. "The goals of the primary school curriculum fall into two related categories, those dealing with the learner and those dealing with mathematical content" (Alberta Education, 1982, p. 4). The program provided for development in problem solving skills, psychomotor skills, and "development and understanding of numeration, operations and properties, measurement, geometry and graphing," as well as fostering a sense of success and accomplishment, [and] positive attitudes towards mathematics and towards learning (p. 4).

The CCF, in spite of being a multi-jurisdictional document, states goals of mathematics education reminiscent of those from Alberta over the decades: for students to "use mathematics confidently to solve problems; communicate and reason mathematically, appreciate and value mathematics; make connections between mathematics and its applications; commit themselves to lifelong learning; [and] become mathematically literate adults, using mathematics to contribute to society" (WNCP, 2006, p. 4).

CURRICULUM AS CULTURAL

Curriculum specifies the cultural knowledge and skills that society values—that society deems most worthwhile to pass on to the new generation. In this way, the curriculum acts to preserve cultural knowledge, as well as to prepare learners for participation in the development of that knowledge. At the same time, curriculum (and in our case the mathematics program of study) is an artifact of society and takes its place as an element of culture. In Albertan culture, and more generally speaking Canadian culture, children and youth study mathematics, they learn to "speak" mathematics and do mathematics. In terms of education, one might argue that mathematics is highly valued. The following quote from 1932 is as true today as it was then.

> From the establishment of the secondary schools to the present time, Mathematics has occupied a prominent place in the curricula. It has taken a position second only to English in the amount of total time given to it. A

programme of studies which did not include a considerable amount of Mathematics would be an anomaly. (Department of Education, 1930, p. 93)

In the elementary school curriculum, number, operations, measurement, and mathematical language and notation were a common theme through the 20th century. In the high school curriculum, algebra, geometry, trigonometry, and analytical geometry were found across the decades. Given the many cycles of curriculum revision over the past 120 years, it is somewhat surprising that the content has remained so stable. Indeed, revisions to the curriculum seem mostly to reflect differences in perspective about the nature of instruction, mathematics, and the child. It seems the public cries for reform have been more about how mathematics is taught and the expectations on what the child should be able to do, rather than changes to the content itself—the knowledge deemed worthwhile.

School Mathematics (Grades K–9)

The mathematics content in the elementary school of the 1940s was "to acquaint the child with the ways in which number is used in our present-day society" (Department of Education, 1940, p. 252). The language of arithmetic was viewed as essential and detailed guidance for how instruction should proceed was specified for the teacher. (For example, see the "tables" in Figure 5.2.) The program of studies emphasized the need for correctness and automaticity in number situations; the application of "concepts, language, generalization and techniques used in mathematical thinking;" an "appreciation of the necessary, invariable and exact relationship involved in mathematical thinking[;] and the necessity of habits of exactness in activities involving number and quantity" (pp. 252–253).

In the 1960s there continued to be an emphasis on the precise use of language, correctness and accuracy, and the use of mathematics to solve problems that arise in everyday living. Mastery was emphasized with little direction given to the teacher for instructional methods. The objectives of the program of studies for elementary mathematics education included "the basic idea" of number, number facts, four foundational operations, decimal number systems, place value, the role of zero, measures, fractions as relationship between integers, estimation of quantitative situations, statistical information through graphs and tables, rapid mental calculation, sound and systematic procedures for problem solving, mathematics vocabulary, ratio, and checking computations to avoid errors (Department of Education, 1963b, p. 24).

The 1978 and 1982 curriculum guides were remarkably different from those that proceeded them. What was typically a few pages of topics and direction for the teacher became pages of objectives (Alberta Education,

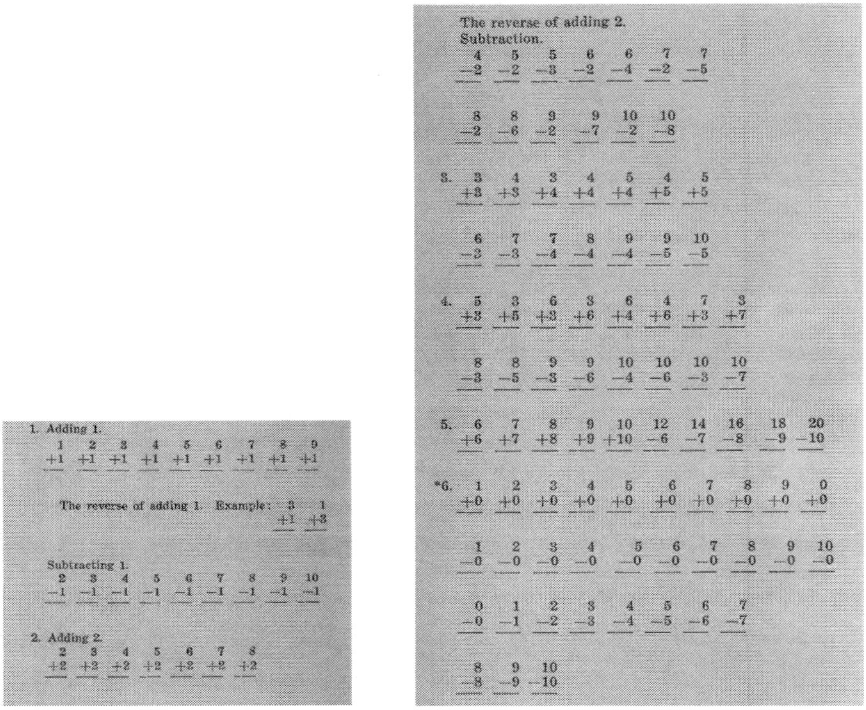

1. Adding 1.
1 2 3 4 5 6 7 8 9
+1 +1 +1 +1 +1 +1 +1 +1 +1

The reverse of adding 1. Example: 3 1
+1 +3

Subtracting 1.
2 3 4 5 6 7 8 9 10
−1 −1 −1 −1 −1 −1 −1 −1 −1

2. Adding 2.
2 3 4 5 6 7 8
+2 +2 +2 +2 +2 +2 +2

The reverse of adding 2.
Subtraction.
4 5 5 6 6 7 7
−2 −2 −3 −2 −4 −2 −5

8 8 9 9 10 10
−2 −6 −2 −7 −2 −8

3. 3 4 3 4 5 4 5
+3 +3 +4 +4 +4 +5 +5

6 7 7 8 9 9 10
−3 −3 −4 −4 −4 −5 −5

4. 5 3 6 3 6 4 7 3
+3 +5 +3 +6 +4 +6 +3 +7

8 8 9 9 10 10 10 10
−3 −5 −3 −6 −4 −6 −3 −7

5. 6 7 8 9 10 12 14 16 18 20
+6 +7 +8 +9 +10 −6 −7 −8 −9 −10

*6. 1 2 3 4 5 6 7 8 9 0
+0 +0 +0 +0 +0 +0 +0 +0 +0 +0

1 2 3 4 5 6 7 8 9 10
−0 −0 −0 −0 −0 −0 −0 −0 −0 −0

0 1 2 3 4 5 6 7
−0 −1 −2 −3 −4 −5 −6 −7

8 9 10
−8 −9 −10

Figure 5.2 An extract from a 1940 curriculum document that recommended the order for teaching number facts (Department of Education, 1940, pp. 260–262).

1978). The 1980s brought an intense focus on the psychology of learning. Although the content topics continued to include number as well as measurement and geometry, problem solving became a major focus of the curriculum. Supplementary curriculum documents produced by Alberta Education were explicit about instruction based on the psychology of learning: children learn by doing, learn by working from the concrete to the abstract, and they need opportunities to maintain and reinforce concepts and skills (Alberta Education, 1980, pp. 35–37).

The contemporary curriculum, derived from the *CCF K–12 Mathematics* (Alberta Education, 2006), is a set of general and specific content outcomes and process outcomes for Grades K–9 (WNCP, 2006) and 10–12 mathematics (WNCP, 2008). Specific outcomes refer to mathematics language, concepts, procedures, representations, and skills. Process outcomes include communication, connections, estimation and mental mathematics, reasoning, problem solving, technology, and visualization. The process outcomes are expected to permeate mathematics lessons, hence students' learning experiences. Consistent with the notion that curriculum draws from the cultural in terms of mathematics that is taught, we see here that the process outcomes reflect the doing of mathematics. Hence, in addition to mathematical knowledge, objects, and concepts as cultural artifact, so too are the processes of mathematics.

High School Mathematics (Grades 10–12)

The secondary high school content has remained remarkably stable over the past century with algebra, geometry, trigonometry, and analytical geometry as the anchors (as illustrated in Table 5.1). Prior to 1960 the content from these topic areas was taught as separate subjects, but since 1960 these mathematical topics have been merged into single courses to create an integrated program in high school mathematics. Since the 1960s, students do not choose from these topics, and they do not study them as a sequence of courses, but rather, they study content from all these topics throughout their high school program. Today "Alberta's math program encourages students to develop mathematical reasoning and problem-solving skills and make connections between mathematics and its applications. Our program also builds students' confidence in their mathematical skills and appreciation of the subject" (Alberta Education, 2016, para. 1).

Resources

In Canadian classrooms, it is standard to find textbooks in the desks and lockers of students. The textbook is a dominant artifact in mathematics education and reflects the mathematics (as cultural legacy) that teachers and children most frequently encounter.

Of particular interest is the relationship between the teacher and the textbook. The Alberta government, through the process of approving resources (textbooks and workbooks) for use in the classroom might be seen as imposing a textbook for mathematics classes. School district leaders and teachers must choose from a small selection (two or three approved resources) for use in the classrooms of Alberta. Having said that, there is a history in Alberta (and other parts of Canada) of teachers contributing to the development of textbooks and other resource materials. Some teachers contribute to the approved textbooks either by working on writing teams or by reviewing the books. Some teachers develop materials locally and circulate them within a school or district as they share their successes and challenges with each other. This should not be surprising as the mathematics textbook is historically a teaching aid designed by teachers specifically

TABLE 5.1 High School Content Across Seven Decades

	1940	1960	1980	2010
Grades 10–12	*Taught as separate courses:* • Algebra • Geometry • Trigonometry and Analytical Geometry • General Mathematics	*Integrated mathematics courses that include:* Geometry, arithmetic, algebra: rules for operations, equations in one and two unknowns, graphs, quadratics, basic operations on polynomials, locus of circle, linear and quadratic functions, polynomials and algebraic equations, rational functions, series, permutations and combinations, mathematical induction, binomial theorem and problem solving	*Integrated mathematics courses:* Number systems, variation, exponents, equations and graphing, geometry, polynomials, trigonometry, presentation of data and descriptive statistics, relations and functions, quadratic functions, equations and applications, coordinate geometry, systems of equations, radicals, quadratic relations, logarithms, sequences and series	*Integrated mathematics courses:* Academic stream includes: • Algebra • Geometry • Logical Reasoning • Mathematics Research Project • Measurement • Number • Permutations, Combinations & Binomial Theorem • Probability • Relations and Functions • Statistics • Trigonometry

Curriculum in Canada ■ 113

for the purpose of exposition of lessons in mathematics. Many Canadian students today will encounter one of only a handful of textbooks produced specifically for the program of studies for a particular jurisdiction. Because the CCF of the WNCP is the basis of the mathematics curriculum in 10 different jurisdictions, and because textbooks were produced specifically for this curriculum, one might speculate that (to the extent a textbook is used in the mathematics classroom) many children in Canada encounter the same lesson on fractions, regardless of where they are attending school.

Textbooks have transformed over the century with advances in printing and access to digital media forms, from print text based to text and image based, and now to interactive texts. In comparing a textbook from 1936 with one from 2009, a number of differences are noted. (See Figures 5.3 and 5.4 for historical and contemporary examples.) The first is the size of the textbook. *Mathematics for Today* (Lazert & Betz) published in 1936 is 366 pages whereas *MathLinks 9* (McAskill et al.) published in 2009 is 506 pages; a two-page spread in the 1936 textbook is the same size as a single page in the 2009 book. Part of the difference in size of the textbooks is that the early 20th century books were text based and judiciously used images. Most frequently those images were small blackline drawings/sketches of mathematical objects (circles, lines, graphs). Today textbooks are full of colour with images (photographs and drawings) of people, places and things, and

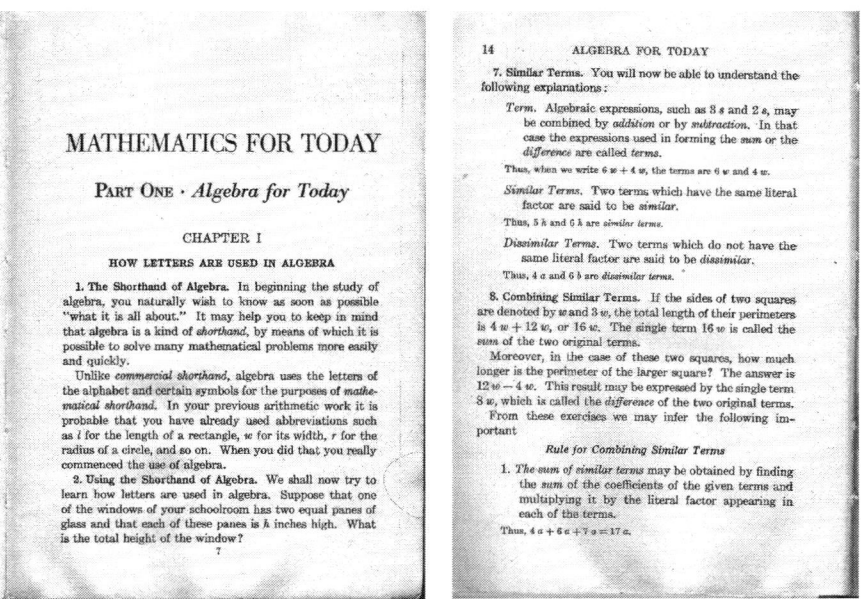

Figure 5.3 Sample from *Mathematics for Today* (Lazert & Betz, 1936).

5.1 The Language of Mathematics

Focus on...

After this lesson, you will be able to...
- use mathematical terminology to describe polynomials
- create a model for a given polynomial expression

algebra
- a branch of mathematics that uses symbols to represent unknown numbers or quantities

The Great Wall is the world's largest human-made structure. It stretches over 6700 km across China. The wall was created by joining several regional walls.

Similarly, mathematics is a developing science made up of several branches, including arithmetic, geometry, and algebra. It is a science that studies quantity, shape, and arrangement. As with any science, mathematics comes with its own unique language. The language of mathematics is universal. It can be understood anywhere in the world.

Look at the following paragraph. How much of it can you read? What languages do you think the paragraph contains?

古代希臘人相信,如果將兩個物體同時拋下,重的那個會先落地。加利略,一個意大利人,做實驗,從比薩斜塔上擲球到地面。實驗顯示出了同的物體運行相同的距離所用的時間相同,與它的質量/重量無關。它們之間的關係是公式 $y=ax^2$ 而不是像許多人預測的 $y=ax$。

Would the algebraic equations be any different if the paragraph was written in any other modern language?

174 MHR • Chapter 5

Figure 5.4a Sample from *MathLinks 9* (McAskill et al., 2009). ©McAskill, Watt, Zarski, Balzarini, Johnson, Kalwarowsky, Licorish, & Webb (2009). *Math Links 9* used with permission of McGraw-Hill Education. *(continued)*

Curriculum in Canada ■ 115

Explore the Language of Algebra

1. **a)** For the algebraic expression $5a + 4b$, what terminology can you use to describe the numbers 4 and 5, and the letters a and b?
   b) What terminology can you use to describe the expression $-7x^2$ and its parts?

2. Make up a real-life situation and write an algebraic expression for it. What do the parts of your expression represent?

3. Algebraic expressions can have different numbers of **terms**.

Number of Terms	Examples
1	5 $7x$ $-3ab$ $\dfrac{y}{2}$
2	$5 + x$ $3x^2 - 2$ $7xy + z^2$
3	$1 - x + y$ $2x^2 - 3x + 5$ $a + b + c$

   term
   - an expression formed from the product of numbers and/or variables
   - $9x$ is a term representing the product of 9 (coefficient) and x (variable)
   - a constant value, such as 5, is also a term

   Write other examples of expressions with one term, two terms, and three terms.

4. A study of algebra includes working with **polynomials**. They are named by their number of terms. How do the names *monomial*, *binomial*, and *trinomial* relate to the number of terms in an expression?

   > What common words do you know that have prefixes of *mono, bi,* and *tri*?

   polynomial
   - an algebraic expression made up of terms connected by the operations of addition or subtraction
   - $3x^2 - 4$ has two terms. $3x^2$ and 4 are connected by the operation of subtraction.

Reflect and Check

5. Look at the algebraic equations in the paragraph written in Chinese on the previous page. Use as much algebraic terminology as you can to describe them.

   Language Link
   The word *algebra* comes from Arabic. The word originated in Iraq. Around A.D. 830, Mohammad al-Khwarizmi of Baghdad wrote a book called *Hisab al-jabr w'al-muqabalah*. This book summarized Hindu understandings of equations and how to solve them. The whole title was too hard for some Europeans so they kept only the word *al-jabr*. We get the term *algebra* from that Arabic word.

   Web Link
   For more information about the history of algebra, go to www.mathlinks9.ca and follow the links.

Figure 5.4b (cont.) Sample from *MathLinks 9*

mathematical objects and representations. The 2009 textbook is dense with full colour photographs, drawings, and visual cues.

More significantly is the pedagogical perspective of the textbook, how does the textbook *teach* the concepts? In the Lazert and Betz text from 1936, the textbook uses the second person and speaks directly to the student, as does the 2009 text. However, the exposition of the lessons is treated differently. Whereas the 1936 text asserts definitions and has students connect to mathematics immediately, the 2009 begins by making an analogy of mathematics being a science made of many branches similar to the Great Wall of China having been made by joining several regional walls. The 1936 text defines *monomial, binomial* and *polynomial*. In contrast, the 2009 text asks students to think where they have heard the prefixes *mono, bi,* and *poly*; not until some examples are provided are the words linked to the number of terms in an expression.

In addition to textbooks, other teaching and learning aids can be found in elementary school classrooms. Base-ten blocks, fraction tiles, geometrical shapes and solids, and measuring tools of various sorts are commonly used in elementary school classrooms. The contemporary curriculum (Alberta Education, 2007/2014) mandates the use of learning aids for some specific outcomes, for example: "Illustrate, concretely and pictorially, the meaning of place value for numerals to 1000" (p. 20) and "Demonstrate with and without concrete materials, an understanding of division (3-digit by 1-digit), and interpret remainders to solve problems" (p. 27).

In the secondary school classroom, scientific calculators and graphing calculators are standard. Indeed, the use of technology is mandated in the program of studies through the process outcomes and is required as textbooks no longer include information such as trigonometric and logarithm tables. Additionally, the graphing calculator is used extensively for graphing functions in the high school mathematics classroom and is required for the diploma examinations. "Calculators are required to be used when writing mathematics and science diploma exams...At minimum, a scientific calculator...when writing science diploma exam...An approved *graphing* calculator (emphasis in original) is required when writing mathematics diploma exams" (Alberta Education, 2015, p. 1). Given the expectation that students will use the graphing calculator on the diploma examinations, it is not surprising that they are fully integrated in the lessons (curriculum) in which learners participate in high school mathematics.

Content and Organization of Contemporary Curriculum in Alberta

The *CCF for K–12 Mathematics* (WNCP, 2006, 2008) is the basis for the current programs of study for mathematics in Alberta. In so much as the CCF is agreed upon by the cooperating jurisdictions, each jurisdiction makes independent decisions to create particular programs of study for their educational system, making adaptations as required. The CCF K–9 is used in its entirety by the Province of Alberta for the programs of study for K–6 and 7–9 mathematics. However, Alberta adapted the 10–12 framework to modify and place particular content in specific course sequences for Alberta youth.

The course sequence for K–9 mathematics is one that all students take as a compulsory program in elementary and junior high school.[5] At the high school level, there are three sequences (tracks or streams) in mathematics from which learners choose (see Figure 5.5). Most learners will enter Mathematics 10C (combined course). At Grade 11 they will select from the "dash one" (–1) sequence intended for students who will study calculus, or the "dash two" (–2) sequence intended for those going to the postsecondary level but who do not need calculus, or the "dash three" (–3) course sequence for students who will enter the workforce directly or through trades. Although most students take courses as illustrated by the horizontal arrows, it is possible they can move down from a dash one course to a dash two if they are not successful or they can move up from dash two to dash one with the successful completion of the dash two. There is a route in the Alberta education system, known as the "Knowledge and Employability" route, for youth who are two to three grade levels below their age-appropriate grade in reading, writing, mathematics, and other levels of achievement and "learn best through experiences that integrate essential and employability skills in occupational contexts" (https://education.alberta.ca/knowledge-and-employability/). This route is not part of the CCF.

The *CCF for K–9 Mathematics* (WNCP, 2006) and *10–12 Mathematics* (WNCP, 2008) is structured in such a way that the nature of mathematics and mathematical processes inform the general and specific outcomes that are organized into four strands or conceptual clusters: Number, Patterns and Relations, Shape and Space, and Statistics and Probability (Figure 5.6). Those four strands reflect the historical content of curriculum in Alberta,

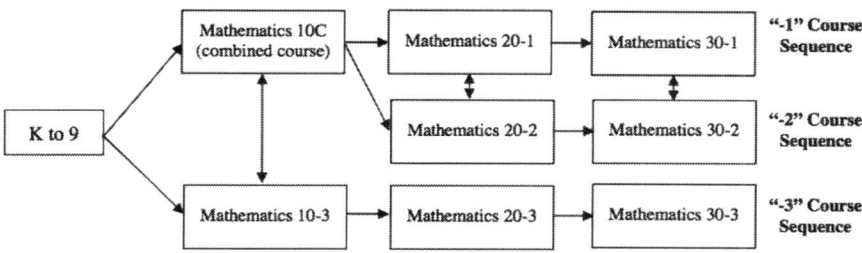

Figure 5.5 High school course sequences. ©Alberta Education. Mathematics Grades 10–12. edmonton, AB (2008) (https://education.alberta.ca/mathematics-10-12/programs-of-study/). Reproduced with permission.

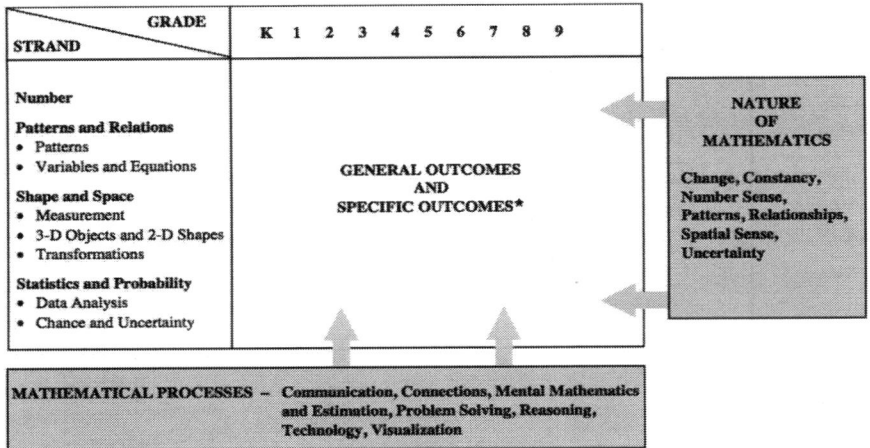

Figure 5.6 *Conceptual Framework for K–9 Mathematics* (Alberta Education, 2007/2016, p. 4). ©Alberta Education. *Mathematics Kindergarten to Grade 9 Program of Studies.* Edmonton, AB. (2007, 2016) (https://education.alberta.ca/media/3115252/2016_k_to_9_math_pos.pdf). Reproduced with permission.

although the strand of statistics and probability is given more emphasis than it had in the first half of the 20th century.

Within the four strands, general and specific outcomes are specified in the programs of study. For example, in the Grade 3 Number Strand, the general outcome is for students to develop number sense (Figure 5.7). Within that general outcome, there are 11 specific outcomes ranging from reciting number sequences to 1000 to demonstrating an understanding of adding and subtracting decimals (limited to hundredths). Together with each of the specific content outcomes are the process outcomes that are to be achieved with that specific content outcome. For example, consider Specific Outcome 3: "Compare and order numbers to 1000 [C, CN, R, V]" (see Table 5.2). The codes (indicating process outcomes) follow the specific outcome and correspond to sections in the front matter of the program of study. These process outcomes are expected to permeate all content, but the use of codes in the outcomes and explaining them in the front matter may result in less attention to them in lessons.

The strand of Shape and Space includes specific outcomes related to recognizing and working with geometric figures and solids to name, measure, derive and use formulas, and create with them. It also includes all forms of measurement and transformations. Extensive work with lines, angles, and triangles is also included in this strand. The Patterns and Relations strand, with its emphasis on patterns, variables, and equations, provides learners with opportunities for generalization and developing means of communicating generalizations and relationships with language (vernacular and

NUMBER

General Outcome
Develop number sense.

Specific Outcomes
1. Say the number sequence 0 to 1000 forward and backward by:
   - 5s, 10s or 100s, using any starting point
   - 3s, using starting points that are multiples of 3
   - 4s, using starting points that are multiples of 4
   - 25s, using starting points that are multiples of 25.
   [C, CN, ME]
2. Represent and describe numbers to 1000, concretely, pictorially and symbolically.
   [C, CN, V]
3. Compare and order numbers to 1000.
   [C, CN, R, V]
4. Estimate quantities less than 1000, using referents.
   [ME, PS, R, V]
5. Illustrate, concretely and pictorially, the meaning of place value for numerals to 1000.
   [C, CN, R, V]

Figure 5.7 Sample from *Program of Study for K–6 Mathematics* (Alberta Education, 2007/2016, p. 20). ©Alberta Education. *Mathematics Kindergarten to Grade 9 Program of Studies.* Edmonton, AB. (2007, 2016) (https://education.alberta.ca/media/3115252/2016_k_to_9_math_pos.pdf). Used with permission.

TABLE 5.2 Process Outcomes and Codes

Communication [C]:	communicate in order to learn and express their understanding
Connections [CN]:	connect mathematical ideas to other concepts in mathematics, to everyday experience and to other disciplines
Mental Mathematics and Estimation [ME]:	calculate and estimate without the use of external memory aids
Problem Solving [P]:	use prior learning in new ways and with new contexts
Reasoning [R]:	develop mathematical reasoning
Technology [T]:	use technology in the learning of mathematics outcomes
Visualization [V]:	develop visualization skills to assist in processing information, making connections and solving problems.

Source: Adapted from Alberta Education (2007/2016, p. 4).

technical), mathematics vocabulary and mathematical notation, structures, and expressions. The Statistics and Probability strand in K–9 programs involves descriptive statistics with basic work in single variable inferential statistics. As well, probability is explored from both experimental and theoretical perspectives. Table 5.3 provides examples of general outcomes with an example of a related specific outcome from a variety of grades.

The high school program of studies for Alberta includes a significant modification to the CCF. Moving away from the organizational structure

TABLE 5.3 Examples of Outcomes from the Grades K–9 Program of Study

Shape and Space	
Grade 2 Measurement	*General Outcome.* Use direct and indirect measures to solve problems. *Specific Outcome.* Demonstrate that changing the orientation of an object does not alter the measures of its attributes. [C, R, V] (p. 68)
Grade 8 Measurement	*General Outcome.* Use direct and indirect measurements to solve problems. *Specific Outcome.* Develop and apply formulas for determining the volume of right rectangular prisms, right triangular prisms and right cylinders. [C, CN, PS, R, T, V] (p. 67)
Patterns and Relations	
Grade 1 Patterns	*General Outcome.* Use patterns to describe the world and to solve problems. *Specific Outcome.* Demonstrate an understanding of repeating patterns (two to four elements) by describing, reproducing; extending; creating patterns using manipulatives, diagrams, sounds and actions. [C, PS, R, V] (p. 60)
Grade 9 Patterns	*General Outcome.* Use patterns to describe the world and to solve problems. *Specific Outcome:* Generalize a pattern arising from a problem-solving context, using a linear equation, and verify by substitution. [C, CN, PS, R, V] (p. 61)
Statistics and Probability	
Grade 5 Chance and Uncertainty	*General Outcome.* Use experimental or theoretical probabilities to represent and solve problems involving uncertainty. *Specific Outcome.* Compare the likelihood of two possible outcomes occurring, using words such as less likely, equally likely and more likely. [C, CN, PS, R] (p. 79)
Grade 7 Chance and Uncertainty	*General Outcome.* Use experimental or theoretical probabilities to represent and solve problems involving uncertainty. *Specific Outcome.* Conduct a probability experiment to compare the theoretical probability (determined using a tree diagram, table or other graphic organizer) and experimental probability of two independent events. [C, PS, R, T] (p. 79)

Source: Adapted from Alberta Education (2007/2014).

of conceptual clusters or strands, Alberta returned to a structure of earlier decades that organized the curriculum by particular topics in the multiple course sequences for high school students. Not only was a long history of multiple course sequences in high school mathematics in Alberta seen as an influence on the decision not to take up the CCF as developed, but stakeholders were also influential.

The content of each course sequence has been based on consultations with mathematics teachers and on the *Western and Northern Canadian Protocol*

(WNCP) Consultation with Post-Secondary Institutions, Business and Industry Regarding Their Requirements for High School Mathematics: Final Report on Findings. (Alberta Education, 2008, p. 10)

Today the high school curriculum in mathematics is divided into three sequences[6] (measurement, algebra and number, relations and functions), each covering three grade levels (10, 11, 12). Additionally, students can take a single course in calculus once they have completed the Grade 12 course in the "dash one" sequence. In Grade 10 most students take the combined course. To provide some sense of the high school curriculum, I provide an abbreviated version of the general and specific outcomes from Math 10C.

- Measurement: Spatial sense and proportional reasoning
  - Solve problems that involve linear measurement (SI and imperial).
  - Apply proportional reasoning in conversions.
  - Solve problems that involve surface area and volume of 3D objects.
  - Develop and apply primary trigonometric ratios with right angle triangles.
- Algebra and Number: Algebraic reasoning and number sense
  - Demonstrate understanding of factors of whole numbers.
  - Demonstrate understanding of irrational numbers.
  - Demonstrate understanding of powers.
  - Demonstrate an understanding of multiplying and factoring polynomial expressions.
- Relations and Functions: Algebraic and graphical reasoning through the study of relations
  - Interpret and explain the relationships among data, graphs, and situations.
  - Demonstrate an understanding of relations and functions.
  - Demonstrate an understanding of slope.
  - Describe and represent linear relations.
  - Determine the characteristics of linear graphs.
  - Relate different algebraic forms of linear relations.
  - Determine equation of linear relation given it in various forms.
  - Represent linear function using function notation.
  - Solve problems that involve linear systems in two variables graphically and algebraically. (Alberta Education, 2008, pp. 13–16)

Students who complete Math 10C then complete either the "dash one" or the "dash two" stream. Topics in the Grade 11 "dash one" sequence include: algebra and number; measurement; relations and functions; trigonometry; and permutations, combinations, and the binomial theorem. Students in Grade 11 "dash two" sequence study geometry, measurement, number and logic, logical reasoning, relations and functions, statistics and probability. Students in the Grade 11 "dash three" sequence study algebra, geometry, measurement, number, statistics and probability. The "dash two" and "dash three" sequences are quite distinct, in spite of being organized around the same topics.

The "dash one" stream is a traditional algebra intense program with extensive work with functions, relations, trigonometry, and polynomials. The "dash two" stream has less emphasis on the traditional algebra content and uses more time to get to functions. Finally, the "dash three" stream is not intended to have depth in algebra as in algebraic structure, but instead uses algebra for practical purposes. To illustrate differences among the course sequences, Table 5.4 outlines the outcomes related to the two content strands of algebra, and functions and relations, in the Grade 11 course sequences.

CURRICULUM AS PERSONAL: THE LIVED CURRICULUM

It is common to think about curriculum in terms of the mandated program of studies and resources (words and ideas fixed in paper and ink or digital files) and forget that there is a lived curriculum (Aoki, 2004/1980)—the encounters and experiences that the human being has with mathematics. Jardine et al. (2008) assert that mathematics is a living discipline. When we think about the curriculum only as a mandated program of studies, we fossilize it and brush aside the human beings who make the curriculum in living it: the teachers who create lessons and in doing so breathe life into the knowledge deemed most worthwhile; the children who contribute to the discipline of mathematics through their learning and active participation in it in school, in the playground, and in their homes and community; and the policy makers who make the decisions about what knowledge is most worthwhile and when in a student's schooling it will be encountered. In this section, the focus turns to the curriculum as brought to life in and by people. The teacher selects from the curricular materials and resources available and sets tasks for learners so they can encounter mathematics. The children experience the mathematics curriculum in the day-to-day activities they engage in that bring forth mathematics. The policy makers decide what mathematics content and processes will comprise the curriculum.

The Policy Makers and the Process of Curriculum Development

I now reflect on how the programs of study come to be what they are and look for the people who are responsible for their shape, their content, and their intentions. The contemporary program of studies for mathematics was

TABLE 5.4 Specific Outcomes Related to General Outcome of Algebraic Reasoning

	Math 20–1	Math 20–2	Math 20–3
Algebra	Develop algebraic reasoning and number sense.		Develop algebraic reasoning.
	Solve problems that involve operations on radicals and radical expressions.	No specific outcomes under the algebra topic	Solve problems that require the manipulation and application of formulas related to: – volume and capacity – surface area – slope and rate of change – simple interest – finance charges.
	Solve problems that involve radical equations.		
	Determine equivalent forms of rational expressions.		
	Perform operations on rational expressions.		Demonstrate an understanding of slope: – as rise over run – as rate of change – by solving problems
	Solve problems that involve rational equations.		
	Demonstrate an understanding of the absolute value of real numbers.		Solve problems by applying proportional reasoning and unit analysis.
Functions and Relations	Develop algebraic and graphical reasoning through the study of relations.		
	Eleven specific outcomes for functions and relations	Demonstrate an understanding of the characteristics of quadratic functions, including: vertex, intercepts, domain and range, and axis of symmetry. Solve problems that involve quadratic equations.	No specific outcomes related to functions and relations

Source: Adapted from Mathematics Grades 10–12 (Alberta Education, 2008).

developed in and by the collaborating jurisdictions in western and northern Canada. In 1996, the first of the multi-lateral curriculum frameworks (WCP, 1996) was mandated in Alberta. The 2007 curriculum is a result of that ongoing collaboration among jurisdictions. Alberta, as the lead province for mathematics, took on the challenge of revising the CCF. Alberta government curriculum designers were guided by the CCF framework developed in 1996 (heavily influenced by the NCTM *Standards* [McCabe, 2000]) and the concerns, challenges and requests from partner jurisdictions and stakeholders.

The ministry curriculum teams worked on the K–6 and 7–9 programs of study, and then the 10–12 programs (guided by the vision and scope for K–12 mathematics). In this way, the K–12 curriculum is conceptualized as a coherent whole. The curriculum makers (who are former teachers with classroom teaching experience) did not work in isolation. The process included ongoing consultation with stakeholder groups consisting of representatives from education (school teachers and education professors), university mathematics faculty, business people, members of industry, and parents (https://education.alberta.ca/mathematics-10-12/). With such an extensive process, curriculum developers were faced with multiple and sometimes opposing demands and expectations of the curriculum by their teaching colleagues, parents, and postsecondary instructors. It was the government curriculum policy makers who negotiated those demands and expectations, so that when the documents went to the minister for approval, the public would be presented with a suitable and acceptable program of study. It is this process of negotiation that reminds us of how political a document the curriculum is.

Teachers as Interpreters and Implementers of Curriculum

The mandated curriculum and curricular resources are mediated through the teacher. It is the teacher who must take stated outcomes and transform them into the lessons where students encounter mathematics. In Alberta, teachers take the programs of study seriously. A recent visit to Alberta by an international scholar brought attention to how commonly teachers in Alberta reference the program of studies when discussing mathematics lessons. Further, teachers are aware that the approved resources for the curriculum have been designed to "cover the content" that is specified in the curriculum (this is an expectation for an approved resource). When the curriculum undergoes changes, teachers can become unsettled. The content and resources with which they are familiar change. Some worry they will have to restart and may not be able to count on their proven lessons, the ones they know work well with their learners. This was the case when the CCF was first brought to Alberta (McCabe, 2000).

The CCF was the first major curriculum change in 40 years. Some teachers saw it as an opportunity, others as a burden. All would have to learn to read a newly structured document, make decisions about sequencing (that in the past had been spelled out for them), and would have to teach mathematics from a "constructivist" perspective. McCabe (2000) describes the situation as follows: "The change mandated by the new curriculum was vast and, for many teachers frightening. The fear of the unknown caused many teachers to react negatively to the new program without truly understanding its philosophy or its purpose" (p. 129). In 2008, when the government indicated it would be looking into the "new" curriculum, teachers were thrown back into the space of the unknown. Many had grown to appreciate the new curriculum, its constructivist foundations, and attention to learners. Teachers had developed new favorite lessons and once again they worried that they would have to start over.

Examining the lived curriculum also points to a realization that teachers, too, are curriculum makers in the local situation, or the small "c" sense of curriculum. They create lessons, tasks, and materials to use with their classes. But their work, like tha- of policy makers, begins to impact teaching and learning beyond their classroom walls when they share their work with teachers in their school or district. Further influence is triggered when province-wide professional organizations, such as the Mathematics Council of the Alberta Teachers Association,[7] provide teachers with a means of sharing their curriculum work more broadly with annual conferences and publications (such as *Delta-K* in Alberta). So, although the curriculum and approved resources are controlled by the government, the lessons and activities that learners encounter are the curriculum work of teachers.

As Experienced by Students

As illustrated earlier, mathematics curriculum has been influenced by the progressive education movement of the early 20th century, the Sputnik response in mid-century, research in the psychology of mathematics learning in the second half of the century, and throughout the past 70 years by the National Council of Teachers of Mathematics. Those influences marked the nature of the encounter children would experience in mathematics in school. Alongside the school-based influences, the experiences of children and youth in and with mathematics includes what is happening in the world around them. The movement of people into cities, the growth of the middle class, and the advances in technology[8] are only a few such out-of-school influences. Coupled with the strong influence of psychology, in particular constructivism and social constructivism, learners' encounters with mathematics have changed over the century. For example, the use of instrumental learning techniques, especially those related to memory, are found in the curriculum from the early 20th century, whereas today children are expected to develop strategies from understanding and generalizing situations. A particularly controversial example from Alberta comes to mind.

In the past two years, a vocal group of parents and some mathematicians have been calling for teachers to get "back to the basics." As part of that demand, the expectation is that children learn the traditional strategies for the four operations and memorize number tables, as they did in the past. It is not clear to which past these advocates refer. However, let us consider the example of tables provided in Figure 5.2 from the 1940s. The child's encounter with mathematics in the 1940s seems to be one in which the child, through repetition, develops mastery of many different techniques for computing the four operations—in this case adding numbers between one and ten. First children would do the sums $1 + 1$, $1 + 2$, $1 + 3$,...and then reverse those sums, $2 + 1$, $3 + 1$, and so on. The next page of the curriculum document goes to a table that demonstrates "double plus one." Presumably, children would work through a number of different strategies documented with tables, memorizing all of them for all cases. Many of the strategies noted in the 1940s document are invoked in the contemporary classroom, but they are developed by the learner as part of their experience of coming to understand number, binary operations, number families, base-ten structure of our number system, the use of their body and environment to reason mathematically. These strategies have become known as "personal strategies," a name that points to the child's understanding and construction. The Grade 1 learner in the class of 2016 experiences those sums much differently than the Grade 1 learner in the class of 1940. Furthermore, children today experience sums quite differently simply by virtue of living in the 21st century (take for example their exposure to children's educational programming on television and the Internet).

It is mathematics—not arithmetic—that has become "basic" for all learners in the 21st century. Insofar as mathematics is a magnificent, complex, and useful human creation, treating it as basic within the aims of education seems reasonable. However, treating mathematics as basic facts is inconsistent with mathematics itself and the ultimate goals of education. Furthermore, treating mathematics as the "basic" gatekeeper to higher education is not in the interest of children and youth. Borrowing loosely from Freudenthal (1973), mathematics is much too important for it to be a gatekeeper.

CURRICULUM AS POLITICAL AND POLITICIZED

What is taken to be "mathematics" has itself fallen pallid and weak, infected with a[n] industrial assembly-line story-line that has trumped its own living ways.
—Freisen & Jardine, 2009, p. 149

In this section, recent events in mathematics education in Alberta are explored to illustrate the political and politicized nature of curriculum. Politicians in their ongoing dance with the electorate have been responding (some would say too hastily) to vocal members of the public and higher education (some would argue minority voices) with calls and demands for changing curriculum. Like many aspects of life today that are infused and directed with messaging in the social, print, and television media, public calls for changes in the mathematics curriculum are constant. Although responsibility for education is placed in the hands of the ministries of education who employ curriculum experts, I wonder who and what is behind the ongoing adjustments that the Alberta mathematics curriculum has been undergoing in the past 20 years. I offer two examples to illustrate the politicized nature of curriculum.

In 2014, the 2012 Programme for International Student Assessment (PISA) scores were made public. Alberta slipped in the ranks. Although still considered a top-performing jurisdiction, Quebec and Ontario both ranked higher than Alberta. The media jumped on the story with headlines like "Alberta's discovery curriculum fails the kids" (Tran-Davies, 2014) and "Math wrath: Parents and teachers demanding a return to basic skills" (Alphonso & Maki, 2014). Since those articles were published, more than 100 articles in the Canadian media have appeared. They do not all espouse the same message; however, it is clear that, through the media, people with strong opposition to the curriculum are voicing their views loudly and attracting the attention of Alberta's Minister of Education.[9] Since these public cries for reform, the Alberta government has made some clarifications to the 2007 program of studies for K–6 mathematics. For example, in the instructional focus for 2007 we read, "By decreasing emphasis on rote calculation, drill and practice and theories of numbers used in paper and pencil calculations, more time is available for concept development" (Alberta Education 2007/2014, p. 4). In the 2014 clarification document, this statement was removed and the following was added: "Learning mathematics includes a balance between understanding, recalling and applying mathematical concepts" (p. 10). Additions were also made to a few specific outcomes, "Understand and apply strategies for addition and subtraction facts to 18. Recall addition and related subtraction facts to 5" (p. 14). These changes came about with a minority of very vocal advocates for memorization of basic number facts. The curriculum makers were able to clarify intentions and tweak outcomes to address what they must have viewed as important critique.

A much more substantial curriculum change occurred when the 2007 curriculum replaced the 1998 program of studies, which included Applied Mathematics. In 1998, when a new course sequence called Applied Mathematics was introduced, it was something very new to teachers, parents, and university registrars (Boyce, 2003; McCabe, 2000). Although the stream was intended to be rigorous, it did not emphasize algebra in its traditional form (including extensive use and practice of algebraic symbolic manipulation), but instead embraced computer software tools such as spreadsheets and graphing calculators for doing computations on equations, linear systems, and other functions and relations. For example, topics such as matrices were added to the applied stream in the context of problem situations, but the matrix operations were manipulated with technology rather than by paper-and-pencil algorithms. The de-emphasis on algebra caused concern for post-secondary institutions, and the Ministry of Education had an uphill battle trying to convince post secondary officials that students would be prepared for university studies in mathematics with the applied course (McCabe, 2000). School counsellors were not sure how to advise students and they agreed with parents that it was best to keep all doors open. By the time the ministry had gained some ground with post-secondary institutions, students who had aspirations for studying at universities stopped registering for the applied stream and pursued the pure mathematics stream. An innovative curriculum that intended to bring authentic investigations and problem solving into the classroom, with the power of technology to do messy computations and difficult visualizations, was discontinued in less than ten years with the introduction of the 2007 Program of Studies in Alberta, returning high school students to the content and methods that had been studied for a century: algebra, analytical geometry, and trigonometry.

From the aims of education discussed in the historical dimension of curriculum section and with these two examples, I have tried to illustrate how mathematics education and curriculum development are both political and politicized. I now conclude this paper by turning to a discussion of education as geographical and situated as I loop back to my opening remarks.

CURRICULUM AS GEOGRAPHICAL AND SITUATED

As I have reinforced many times, education is a provincial responsibility. Hence curriculum, by law, is geographical (and geopolitical). I point to two factors that were notable in triggering the joint initiative of the WNCP: population movement and population density. Canada is a large country and the population is not evenly distributed. The population is centered largely in Ontario and Quebec and the lower mainland of British Columbia. The provinces in the west and the north have relatively small populations. Across the country, employment opportunities do not correspond directly with availability of workers. Not surprisingly, there is movement of children within the country from one jurisdiction to another. Prior to 1996, when children moved from schools in one province to another (Manitoba to Alberta, for example), they would be faced with having to catch up on content they might have missed or repeating content they had already studied. Ministries of Education in Canada take responsibility for programs of study, but they do not take responsibility for producing and publishing curriculum resources (textbooks, in particular). Prior to 1996, teachers and learners in Alberta were faced with textbooks that were not written for their curriculum. Curriculum resources that fit the Alberta programs of study were not available (in part) because the population of the western provinces is relatively small (in Alberta just over 4 million people) and publishers had little incentive to produce materials for small markets. The problem of having children's education disrupted by moving among jurisdictions, with different scope and sequencing of the curriculum, coupled with the desire to create a market large enough to entice publishers to create texts specifically for Western Canada, contributed to the decision of the ministers of education to create the Western Canada Protocol for Basic Education.

I wrote a commentary that appeared in a recent issue of the *Canadian Journal of Science, Mathematics and Technology Education*, "What is a Canadian Mathematics Education?" (Simmt, 2015). I concluded that it was a national discourse community that has emerged in spite of our geographical vastness (and today I add geopolitical diversity). The west and the north, together with the Atlantic provinces, have made significant contributions to bringing together the stakeholders in mathematics education in their desire to create curriculum that illustrates common beliefs about mathematics, teaching, and learning and shared goals of education.

CONCLUDING REMARKS: CURRICULUM AS FRACTAL

Curriculum as mandated, planned and implemented has a fractal footprint. As we examine it as a historical phenomenon, we find elements of it as cultural, political, geographical and personal. But our fractal lens suggests each of those must be examined in their own right. In this paper, I explored how the curriculum children and youth experience in the contemporary classroom in Western, Northern, and Atlantic Canada is developed and transforms over time and with various influences.

The history of mathematics curriculum in Canada continues to be written. It will be influenced by changes in society, advocates who shape the national discourse, new research, new mathematics, and new ways of being in the world. For a century, we have witnessed relatively stable content in mathematics. The technological advances are so swift, so profound, and so permeating that it would be surprising for mathematics curriculum not to take on quite a new shape in the next century. Because of these technological advances, Canada will need people with deep understanding of this content, stronger mathematical reasoning, stronger computational thinking skills, and the capacity to work on hard problems. What might this mean for Canada? I suspect collaboration and cooperation for curriculum renewal is in our future.

NOTES

1. For example, in Alberta the influence of the National Council of Teachers of Mathematics is noted as far back as the 1940s with references to their publications in provincial curriculum documents.

2. Before that, in 1996 the Western Canada Protocol (WCP)
3. O'Shea (2003) provides an illustration of this cyclical process of curriculum reform in an extensive summary of the history of mathematics education in British Columbia.
4. Alberta Department of Education, Alberta Education, Department of Education and Alberta Learning are all names referring to the same body but at different historical periods. Each reference and citation is given in its historically accurate phrasing.
5. For the most part all children attend classes with their age cohort (including children and youth with physical, psychological, and cognitive challenges). Teachers are expected to write educational plans and deliver lessons for learners who are not working at grade level in particular areas, all within the context of their classroom. There are exceptions and some schools provide special services and programs to meet specific needs of learners. The Alberta system works by providing funds to school districts for classroom support when children and youth need exceptional services.
6. Knowledge and Employability is the fourth strand, but it is not available to all students.
7. MCATA is not unique, there are other such organizations in Canada. See (Simmt, 2015).
8. In Canada, more than 80% of households have access to the Internet at home (Statistics Canada, 2013).
9. The public outcry also included a rally on the legislature by 200 people and an online petition signed by 0.4% of the population.

REFERENCES

Alberta Education. (1978). *Program of studies for junior high school mathematics*. Edmonton, Alberta: Author.
Alberta Education. (1980). *Mathematics 10 and 13 interim curriculum guide*. Edmonton, Alberta: Author.
Alberta Education. (1982). *Elementary mathematics curriculum guide 1982*. Alberta, Canada: Author.
Alberta Education. (2006). *WNCP CCF for K-9 Mathematics*. Alberta, Canada: Author.
Alberta Education. (2007/2016). *Mathematics Kindergarten to Grade 9 Program of Studies*. Edmonton, AB. Retrieved from https://education.alberta.ca/media/3115252/2016_k_to_9_math_pos.pdf
Alberta Education. (2008). *Mathematics grades 10–12*. Alberta, Canada: Author. Retrieved by https://education.alberta.ca/mathematics-10-12/programs-of-study/
Alberta Education. (2015). *Writing diploma exams using calculators*. Edmonton, Alberta: Author. Retrieved from https://education.alberta.ca/media/1089192/06-dip-gib-2015-16-using-calculators.pdf
Alberta Education. (2016). *Programs of study (high school)*. Retrieved from https://education.alberta.ca/mathematics-10-12/programs-of-study/
Alphonso, C., & Maki, A. (2014). Math wrath: Parents and teachers demanding a return to basic skills. *Globe and Mail*, January 7. Retrieved from http://

www.theglobeandmail.com/news/national/education/petitions-press-provinces-to-put-emphasis-on-basic-math-skills/article16240118/

Aoki, T. (2004 /1980). Toward curriculum inquiry in a new key (1978/1980). In B. Pinar & R. Irwin (Eds.), *Curriculum in a new key* (pp. 89–110). Mahwah, NJ: Lawrence Erlbaum Associates.

Boyce, S. (2003). *When will I use this?* Unpublished Master of Education thesis. Edmonton, AB: University of Alberta.

Council of Ministers of Education, Canada [CMEC]. (2008). *Education in Canada: An overview*. Retrieved from http://www.cmec.ca/299/Education-in-Canada-An-Overview/

Davis, B. (1996). *Teaching mathematics: Toward a sound alternative*. New York, NY: Routledge.

Department of Education. (1930). *Handbook for secondary schools Alberta*. Edmonton, AB: W. D. McLean King's Printer.

Department of Education. (1940). *Programme of studies for the elementary school: Grades I – VI*. Edmonton, Alberta: Author.

Department of Education. (1963a). *Senior high school curriculum guide for mathematics*. Edmonton, Alberta: Author.

Department of Education. (1963b). *Program of studies for elementary schools of Alberta*. Edmonton, Alberta: Author.

Freudenthal, H. (1973). *Mathematics as an educational task*. Dortrecht, the Netherlands: D. Reidel.

Friesen, S., & Jardine, D. (2009). On field(ing) knowledge. In B. Sriraman & S. Goodchild (Eds.), *Relatively and philosophically earnest: Festschrift in honor of Paul Ernest's 65th birthday* (pp. 147–172). Charlotte, NC: Information Age.

Jardine, D., Clifford, P., & Friesen, S. (2008). *Back to the basics of teaching and learning*. New York, NY: Routledge.

Lazert, M. E., & Betz, W. (1936). *Mathematics for today*. Toronto, Canada: Ginn and Company.

McAskill, B., Watt, W., Zarski, C., Balzarini, E., Johnson, B., Kalwarowsky, E., Licorish, T., & Webb, M. (2009). *MathLinks 9*. Toronto, Canada: McGraw-Hill.

McCabe, K. (2000). *Implementation of the WCP in mathematics in AB*. Unpublished Master of Education thesis. Edmonton, AB: University of Alberta.

O'Shea, T. (2003). The Canadian mathematics curriculum from the new math to the NCTM Standards. In G. Stanic & J. Kilpatrick (Eds.), *A history of school mathematics* (Vol. 1. pp. 843–896). Reston, VA: National Council of Teachers of Mathematics.

Simmt, E. (2015). What is a Canadian mathematics education? A national discourse community. *Canadian Journal of Science, Mathematics and Technology Education, 15*(4), 407–417.

Statistics Canada. (2013). *Canadian Internet use survey, 2012*. Retrieved from http://www.statcan.gc.ca/daily-quotidien/131126/dq131126d-eng.htm

Tran-Davies, N. (2014). Alberta's discovery curriculum fails the kids. *Globe and Mail*, March 10. Retrieved from http://www.theglobeandmail.com/news/national/education/alberta-education-reforms-ignore-kids-parents/article17390684/

Western Canada Protocol [WCP]. (1996). *The common curriculum framework for K-12 mathematics: Grade 10 to Grade 12*. Western Canada Protocol for Collaboration in Basic Education.

Western and Northern Canadian Protocol [WNCP]. (2006). *The common curriculum framework for K–9 mathematics*. Western and Northern Canadian Protocol for Collaboration in Education.

Western and Northern Canadian Protocol [WNCP]. (2008). *The common curriculum framework for Grades 10–12 mathematics*. Western and Northern Canadian Protocol for Collaboration in Education.

CHAPTER 6

MATHEMATICS CURRICULUM IN THE UNITED STATES

New Challenges and Opportunities

Janine Remillard
University of Pennsylvania

Luke Reinke
University of North Carolina at Charlotte

The mathematics curriculum in the United States is difficult to characterize. Curriculum decisions in the United States are made at multiple levels of a complex system. These decisions are further complicated by a shifting terrain of curriculum resources, propelled forward by the Internet and globalization. Understanding the mathematics curriculum in the United States requires understanding how curriculum decisions are made at different levels of the system and, ultimately, how these decisions shape students' opportunities to learn. A primary aim of this chapter is to describe this process in broad strokes, using the contemporary reform initiative referred to as the Common Core State Standards, begun in 2008, as an illustration.

Partially responsible for the complexity is the fact that the United States does not have a single school system. The United States is a federation of 50 states, in which governing powers are distributed, some to the states and some to the federal government, as specified by the Constitution. Governance over education is the responsibility of the states, which typically grant considerable autonomy to local school districts and their elected governing boards. There are more than 18,000 local school districts, which serve over 50 million students, kindergarten through Grade 12 (ages 5 through 18), in over 90,000 schools.[1] Still, as we discuss in the chapter, the federal government uses a variety of means to exert influence on educational policy.

Our discussion of *curriculum* is informed by the conceptual framework proposed by Remillard and Heck (2014), which characterizes the curriculum policy, design, and enactment system. Building on previous curriculum implementation frameworks and modified to accommodate contemporary phenomena influencing education, such as increased accountability, globalization, and connectivity via the Internet, this framework situates the curriculum enactment process within a complex system of policy and practice (Figure 6.1). Briefly, this framework acknowledges that curriculum making occurs at multiple levels and in different forms. The *official* arena refers to the aims specified by officially authorized agencies and the consequential assessments used to measure outcomes. The official arena also includes the *designated curriculum*, which includes instructional plans specified by local governing authorities, and often includes the identification of instructional resources. The *operational* arena includes the curriculum designed and

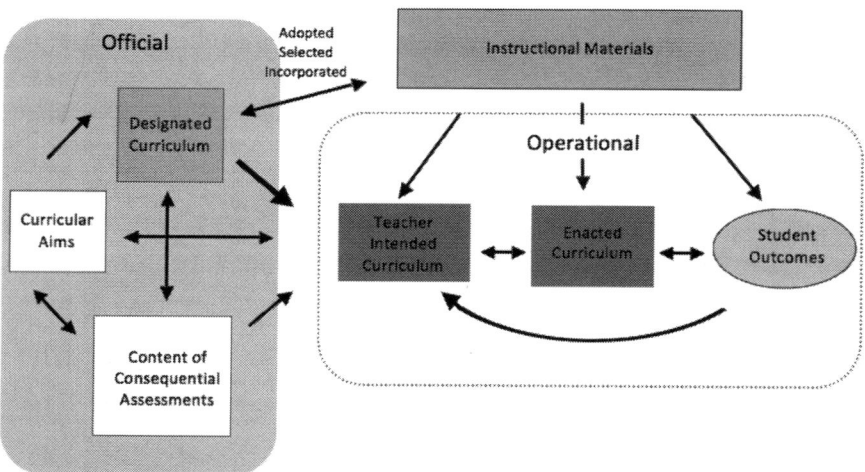

Figure 6.1 Curriculum policy, design, and enactment conceptual framework. *Source:* Remillard and Heck, 2014, p. 709.

enacted by teachers and with students. Relevant aspects of this framework are described throughout the chapter, as needed.

This curriculum policy, design, and enactment framework assumes a definition of the mathematics curriculum, which we also adopt in this chapter: "a plan for the experiences that learners will encounter, as well as the actual experiences they do encounter, that are designed to help them reach specified mathematics objectives" (Remillard & Heck, 2014, p. 707). This chapter begins by describing the official curriculum, from both a national and state perspective, and then discusses the development and use of instructional resources to consider how curriculum is operationalized locally.

U.S. CURRICULUM POLICY AND PRACTICE AT THE NATIONAL LEVEL

Building on the best of existing state standards, the Common Core State Standards provide clear and consistent learning goals to help prepare students for college, career, and life.

—Common Core State Standards Initiative, n.d.a., para. 1

A Standards Story

In 2008, the National Governors Association (NGA), the Council of Chief State School Officers (CCSSO), and Achieve, Inc. released a report entitled *Benchmarking for Success: Ensuring all U.S. Students Receive a World-Class Education* (National Governors Association, the Council of Chief State School Officers, & Achieve, Inc., 2008). The report, which was initiated by state governors and multiple state education leaders and funded by two private foundations with ties to large corporations in the United States (the Bill & Melinda Gates Foundation and the General Electric [GE] Foundation), highlighted an expanding knowledge economy, which demands "knowledge-fueled innovation," problem solving, and communication. Using comparisons to other countries in educational access and outcomes, the authors assert: "The United States is falling behind other countries in the resource that matters most in the new global economy: human capital" (p. 5).

The report proposed a five-step plan for "Building Globally Competitive Education Systems" (NGA et al., 2008, p. 23), which begins with "adopting a common core of internationally benchmarked standards in math and language arts for grades K–12" (p. 24). Steps 2 through 5 offer somewhat of a blueprint for implementing these standards in the decentralized U.S. education system. They include leveraging collective influence on developers of instructional materials and assessments (Step

2), revising state policies related to recruiting, preparing, and supporting teachers (Step 3), holding school systems accountable for making progress with respect to "international best practices" (Step 4), and measuring state-level performance in student achievement and attainment "in an international context" (Step 5; p. 26–34).

An Attempt at Common Standards

The widely discussed *Common Core State Standards* (CCSS), rolled out in 2010, represented the first step of the five-step plan proposed by the *Benchmarking for Success* report. A team of two mathematicians and a mathematics educator were charged with primary authorship of the standards for mathematics (CCSSM) in 2009.[2] Under an ambitious six-month timeline, drafts of the grade-by-grade content goals for kindergarten through high school were reviewed by stakeholders and a validating committee, and revised, before being released to the public in early 2010.[3]

A major goal of the CCSS initiative was to develop a single set of standards that could be voluntarily adopted by a majority of U.S. states, reducing the inconsistency in curriculum expectations across the country (McCallum, 2015) and allowing curriculum designers to focus on a single set of standards as opposed to separate standards for each state. Further, the *CCSS for Mathematics* were described by advocates as having three characteristics essential to the aim of international benchmarking: *focus, coherence, and rigor*. The importance of these characteristics emerged through analyses of the curriculum frameworks of the highest performing countries in international comparisons (e.g., Trends in International Mathematics and Science Study [TIMSS] and Programme for International Student Assessment [PISA]), and were found to be largely absent from most state standards in the United States (Schmidt, Wang, & McKnight, 2005).

Focus, according to standards author William McCallum (2015), "means attending to fewer topics in greater depth at any given grade level, giving teachers and students time to complete that grade's learning" (p. 6). In other words, the list of specific standards to be addressed at any given grade level was shorter than had previously been typical.

Coherence was used to describe the way the standards develop over time, within a grade level and across grades. Rather than representing a list of discrete learning outcomes, the standards build in a logical sequence, one that reflects both mathematical structures and implications of research on cognitive development in general and learning in specific domains.

Rigor was defined as "balancing conceptual understanding, procedural fluency, and meaningful applications of mathematics" (McCallum, 2015). This attempt at balance was evident in the range of verbs used to describe specific learning outcomes, including *use, interpret, understand, determine, solve*, and *apply*.

The designers also included a set of *Standards for Mathematical Practice,* which describe habits of mind or ways of working necessary for "mathematical proficiency," such as "Mak[ing] sense of problems and persever[ing] in solving them" and "Look[ing] for and mak[ing] use of structure" (NGACBP & CCSSO, 2010). The Common Core State Standards Initiative (n.d.b.) attributed these practices to longstanding values in mathematics education, represented by the five process standards identified by the National Council of Teachers of Mathematics [NCTM] (2000) and the five intertwining strands of mathematical proficiency, included in the National Research Council (NRC) report *Adding It Up* (NRC, 2001).

Approaches to Implementation: Carrots and Sticks

Efforts in pursuit of Steps 2 through 5 of the five-step benchmarking plan involved a confluence of initiatives, aimed at providing states with incentives, monitoring tools, and supports. These initiatives, and accompanying funds, emerged from public and private sectors, including federal and state agencies, nongovernmental organizations, and the private sector.

To incentivize states to participate in the Common Core initiative, the U.S. Department of Education (ED), under President Obama, incorporated these standards into a major funding initiative aimed at encouraging state-level innovation in K–12 education. The Race to the Top (RTTT) competition required applicants to adopt statewide standards, measures of educator effectiveness, and other school reform policies aimed at school improvement. Once the CCSS were released in June of 2010, ED invited the 36 states being considered for RTTT grants to amend their proposals "regarding the adoption of common college- and career-ready standards by August 2, 2010" (U.S. Department of Education, n.d.). RTTT did not require adoption of the CCSS, specifically, but by August, 2010, 45 states and the District of Columbia had officially adopted these standards, with plans to fully implement them within three years.

New Assessments

There was general agreement that appropriate assessment and accountability systems were essential to move U.S. schools toward these new standards. Under previous policy, states had developed their own standards and used their own tests to measure attainment; arguably, none of the existing standards or assessments was aligned with the Common Core in terms of content or rigor. Promoters of the CCSS initiative argued that common standards would allow states to wield collective influence over developers of assessments, presumably contributing to their quality.

In 2010, ED funded two assessment consortia to develop assessments aligned with the CCSS. States were invited to join either consortium and most did. The Partnership for Assessment of Readiness for College and

Careers (PARCC) began with 19 member states (plus 7 participating states) and Smarter Balanced Assessment Consortium (SBAC) began with 21 member states (plus 9 participating states). Member states oversaw the development of the assessments, and along with participating states and territories pledged to use the completed assessments beginning in the 2015–2016 school year.

Curriculum Materials

In the United States, as in countries around the world (Valverde, Bianchi, Wolfe, Schmidt, & Houang, 2002), textbooks and other instructional materials are used in most mathematics classrooms and influence what is taught (Chingos & Whitehurst, 2012; Schmidt, Houang, & Cogan, 2002). They also serve as tools to ensure coherence across schools and provide teachers with support and guidance (Ball & Cohen, 1996; Remillard & Taton, 2015).

The textbook publishing industry in the United States is a large market-driven enterprise that relies on schools purchasing their products. Anticipating the potential role that curriculum publishers might play in supporting or impeding uptake of the Common Core standards, those overseeing development of the standards charged the writing team to produce guidelines for publishers. The Publishers' Criteria for the Common Core for Mathematics documents, one for K–8 and another for high school, became available in 2013.[4] They included a list of 10 or 8 criteria, respectively, and a list of general indicators of quality of instructional materials and tools for mathematics. "States, districts, and publishers," the introductions of both documents argued, "can use these criteria to develop, evaluate, or purchase aligned materials, or to supplement or modify existing materials to remedy weaknesses" (NGA, CCSSO, Achieve, Council of the Great City Schools, National Association of State Boards of Education, 2013a, 2013b).

Publishers did not wait for the criteria before beginning to revise old editions of textbooks and develop new programs. The release and rapid adoption of the CCSS led to a storm of activity inside and outside of the mainstream publishing industry. Many publishers were quick to adorn their materials with insignia claiming, *Common Core Aligned.*

The need for new instructional materials, engendered by the standards, occurred at an unprecedented period in the educational materials market. The increasing ease with which digital content could be produced and disseminated through the Internet created opportunities for many would-be developers of instructional materials and tools. Private foundations also contributed to this flurry of activity, providing funding for development of curriculum resources or creating Internet-based platforms for networks of teachers to share their own materials (e.g., http://k12oercollaborative.org/). See

TABLE 6.1 Timeline of Major Events in the Development and Rollout of the CCSS

2008	NGA, CCSSO and Achieve released *Benchmarking for Success: Ensuring all U.S. Students Receive a World-Class Education*, proposing a 5-step plan for "Building Globally Competitive Education Systems," beginning with "adopting a common core of internationally benchmarked standards in math and language arts for grades K–12."
2009	Common standards in mathematics and language arts were developed, reviewed, and made available for public comment.
2010	After state review, final Common Core State Standards and validation report were released. Adopting statewide, college-and career-ready standards was tied to Race to the Top funding. Department of Education funded PARCC and SBAC assessment consortia.
2011-12	45 states and the District of Columbia officially adopted the CCSSM, with plans to implement in three years. (A number of states later withdrew, discussed later.)
2013	The Publishers' Criteria for the Common Core for Mathematics was released to guide the development of CCSS-aligned materials for K–8 and High School.
2015	First official administration of PARCC and SBAC. (Many states chose not to administer either test.)

Table 6.1 for a timeline of major events in the development and rollout of the CCSS.

Curriculum Making in the United States

This story of the development and rollout of the CCSSM serves two purposes: First, it provides a brief introduction to the most recent chapter (as of 2016) of national curriculum reform in the United States; second, and more importantly, it illustrates the complexities and idiosyncrasies of the U.S. educational systems. In the following discussion, we explain some of these unique characteristics, showing their impact on the most recent efforts to influence the mathematics curriculum. Throughout, we use the CCSSM initiatives as a primary example, making comparisons to the curriculum reforms associated with the NCTM *Standards* (NCTM, 1989) in the 1990s and to the No Child Left Behind policies in the early 2000s. Comparing these recent reform efforts highlights the complexity of curriculum making in the United States and illustrates the effects of recent social evolutions on the process, including a growing global economy and Internet connectivity. We conclude with a discussion of critiques of national curriculum reforms in the United States, which further illustrates these issues.

Local Control

Local control is highly valued in the United States. Although states are responsible for educational governance, local school districts are granted considerable autonomy. Local school districts are expected to be responsive to their local constituencies, especially parents. Under this arrangement, states, not a federal agency, specify target learning outcomes and determine how districts will be held accountable for meeting them—most often with a common assessment. School districts generally decide how to support and guide instruction in their schools, including which materials to adopt. Some funding for education comes from the state, but a substantial amount of funding is raised locally, typically through property and other taxes. (Distribution of curricular decision making authority is summarized in Table 6.2.)

Districts vary in size; the New York City Department of Education, for example, serves a population of over one million students (National Center for Educational Statistics, n.d.). Other districts are considerably smaller and may serve several hundred or several thousand students. More recently, individual schools, which have been granted charters from their home state, can operate independently from districts and may serve fewer than ten students.[5]

Federal Influence

Over the years, the U.S. federal government has sought, with some success, to establish influence over educational policy through indirect mechanisms, such as providing targeted funds for programs and research, and by increasing the prominence of the education arm of the federal government.

TABLE 6.2 Distribution of Curricular Decision-Making Authority Across Three Levels of Governance

Federal	☐ Possesses no authority in matters of curriculum ☐ Provides funding to the states for targeted educational initiatives ☐ Influences educational policy using conditional funding
State	☐ Possesses authority for creating educational policy; some state constitutions delegate most educational policy to districts ☐ Determines curriculum standards and summative assessments ☐ Some states identify a list of approved instructional materials
District/School	☐ Selects and purchases instructional materials to make available for teachers ☐ Determines policy for use of instructional materials

In 1979, the federal Department of Education was established and given cabinet-level status, expanding the authority of the previous Office of Education, established in 1867. Still, debates continue to rage across the United States over the role of the federal government in education and what some refer to as "government overreach."

A far-reaching example of the use of targeted funds is the Elementary and Secondary Education Act (ESEA), passed in 1965, under the Johnson administration. The ESEA included provisions that distributed funding to schools and districts with a high percentage of students from low-income families. Although states were given leeway over how to use the funds, receiving them required compliance with federal mandates. The ESEA has been reauthorized by the U.S. Congress about every five years, often expanding the reach of federal policy.

The G. W. Bush administration's 2001 reauthorization of the ESEA, under the title No Child Left Behind (NCLB), illustrates the use of funding mechanisms to expand federal reach over educational policy. NCLB required states wishing to receive supplemental federal funds for schools serving disadvantaged students to comply with a number of accountability measures. State officials were required to set rigorous standards in mathematics and English language arts, measure student achievement, and impose sanctions on schools not making *adequate* progress. The result: States adopted standards or revised old ones. But these standards, and the states' assessment programs, varied in consequential ways. An analysis of the mathematics learning goals for K–8, outlined in state standards available in 2005, found substantial variation in (a) the grade placement of particular expectations, (b) the language used to convey learning (e.g., understand, explore, memorize), and (c) the level of specificity used to indicate *grain size* (see Reys, 2006). For many policy makers, educational advocates, and curriculum developers, this incoherence was a major weakness of the U.S. curriculum (Schmidt et al., 2002).

The Open Marketplace

The United States was founded on free-market and democratic principles. Likewise, public education continues to be open to engagement from the private sector, both in terms of products and ideas. Historically, the educational publishing industry has provided the most prominent example of market influence in education. Textbooks and other materials, marketed by national or international corporations and purchased by school districts, continue to play an influential role in curriculum policy and enactment. The digital revolution has rapidly expanded the types of educational resources available to schools to include learning management systems, adaptive instructional materials and assessments, and interactive tools. The Internet has also made it possible for independent developers of instructional resources, including teachers, to market their products directly to schools or teachers, bypassing any formal adoption process implemented by a state or school district.

Schools in the United States are also open to ideas and innovations from the private sector and independent organizations. Across the states, many local and national organizations have a stake in education and wield influence in a number of ways. The NCTM, a large professional organization of mathematics educators, is one such example. NCTM led the development of the first set of curriculum standards in the United States, *The Curriculum and Evaluation Standards for School Mathematics* (NCTM, 1989). Other influential organizations include the NGA, CCSSO, and Achieve, all of which played a central role in instigating the CCSS initiative. Private foundations, like Gates or GE, which have philanthropic missions and corporate interests, also influence educational policy through funding decisions.

Curriculum Reform by Coalition

Both the NCTM *Standards* and the CCSS movements were initiated by professional or advocacy organizations; that is, they were not developed within federal or state governmental processes. But in order to affect measurable change, they needed to leverage support and participation of stakeholders within the public and private sectors. The NCTM *Standards*, released in 1989, called for a substantial shift in the school mathematics curriculum that included a decreased emphasis on rote procedures taught in isolation and an increased emphasis on mathematical reasoning, conceptual understanding, and problem solving in realistic contexts. At the time, not all states had curriculum frameworks, but many had NCTM members in their state departments of education, who influenced the uptake of the standards at the local level (CCSSO, 1995). Cognizant that instructional materials would be critical to supporting standards implementation across the United States, the National Science Foundation (NSF), an independent federal agency that issues limited-term grants, funded a number of instructional material development projects, beginning in 1991. The resulting 15 programs, often referred to as Standards-based, were eventually marketed by commercial publishers, as was required by the NSF (Senk & Thompson, 2003). Testing and accountability were not part of the national landscape at the time.

The CCSS initiative was driven forward by a somewhat different coalition of groups. As described at the beginning of this chapter, the coalition behind the CCSS initiative represented state government leaders and private educational advocates and foundations, many with ties to large corporations. The NCTM did not play a steering role, but offered support as the initiative began to pick up speed; NCTM's influence in the process is evidenced, many would argue, in the emphasis given to the Standards for Mathematical Practice. Given the accountability system already in place, due to NCLB, the ability to measure student attainment of the goals articulated by the CCSS was seen as critical to ensuring implementation. ED funded the development of new assessments and provided incentive for states to adopt the new standards and use the new tests through RTTT.

As with the NCTM *Standards*, the availability of aligned curriculum materials was generally understood to be a necessary cog in the implementation process of the CCSSM. Commercial publishers, educational researchers, mathematicians, and others were quick to answer the call for *aligned* materials. The wide availability of instructional materials and tools, by way of the Internet, from which teachers and schools could select resources, represents one difference between the CCSS initiative and the NCTM *Standards* movement.

The Direction of the Mathematics Curriculum: A Source of Ongoing Debate

Although curriculum reforms in the United States tend to be propelled forward by the hands of many, they are never uniformly embraced by all Americans and they tend to encounter critique and dissent. The questions at the center of these debates, at least in the most recent decades, tend to be about the ultimate aims of mathematics education (and education more broadly), what mathematics knowledge is most needed and to what ends, and how mathematics should be taught. In the most recent reform movement, the debates about the content of the CCSS are caught up in national and local battles over who should control curriculum decisions.

The most vocal critics of the NCTM curriculum reform movement in the 1990s were mathematicians and parents, who argued that the approach to teaching math promoted by the NCTM *Standards* and related curriculum programs placed too much emphasis on conceptual development and student generation of strategies at the cost of instruction on clearly established skills and conventions that laid the foundation for higher mathematics. Critics labeled the emphasis on problem solving and multiple solution strategies *fuzzy math*, and encouraged parents to challenge schools that adopted NSF-funded curriculum programs, inciting what came to be called the "math wars" (Schoenfeld, 2004).[6] A quieter critique of the NCTM *Standards* movement came from critical educators who questioned the *mathematics for all* rhetoric of the reform as "empty promises and sloganizing" (Martin, 2003, p 10). Martin (2003), along with Apple (1992) and other scholars, observed that despite their call for broadening access to mathematics, the NCTM *Standards* did little to shift ideas about official knowledge.

In many ways, the mathematical focus of the CCSSM, and specifically the three-part definition of rigor, coherence, and focus, reflected an attempt to stake out a middle ground between the two extreme camps in the math wars that surrounded the NCTM *Standards* reforms. Critics of the content of the CCSSM exist, however. The lack of emphasis on statistics, particularly in the early grades, was questioned by educators who believed fluency with data is critical to career readiness in the 21st century (Franklin, 2013). Others raised questions about the developmental appropriateness of the standards for younger grades (Chicago Teachers Union, 2014). Some parents took to the Internet to rail against nontraditional approaches to teaching multi-digit computation, attributing (often inaccurately) their presence in their children's curriculum to the CCSSM.

The most vociferous condemnations of the initiative, which came from both inside and outside of the mathematics education community, challenged the extent to which corporations and private funders were allowed to control the school curriculum, particularly when many of the same corporations were selling instructional materials, equipment, tests, or other services to school districts (Wolfmeyer, 2014). A second but related critique was in the way the school curriculum was being used to prepare a corporate workforce. The CCSS initiative, the argument went, positioned American school children as workers, rather than problem solvers, decision makers, social critics, and change agents (Chicago Teachers Union, 2014; Wolfmeyer, 2014). An equally vocal criticism of the CCSS initiative was leveled at the testing regime that accompanied the rollout of the standards. The NCLB reforms had ushered an unprecedented emphasis on test-driven curriculum into U.S. schools. The threat of new tests that had high stakes for teachers and students led to an "opt-out of testing" movement. The CCSS initiative, because of its aim of forging a common set of standards across the United States, also faced challenges by advocates of states' rights and district autonomy, issues discussed in the following section.

Mathematics Curriculum Reform in the United States

Two characteristics emerge from the above description of governing structures and recent curriculum reform in the United States. First, the federal government is not able to drive centralized curriculum reform, but plays a supporting and enabling role. Unlike many other countries, curriculum revisions do not occur on a routine cycle, but in response to perceived national crises. Independent organizations and for-profit companies tend to initiate these reforms and propel them forward. As a result, when curriculum reforms occur, they are guided by a coalition, as opposed to a top-down process often seen in countries where a national ministry of education sits at the helm. Second, the curriculum-by-coalition approach is further buoyed by the open-market predilections of the U.S. school system. Consequently, curriculum reform messages come from multiple sources, including standards documents, curriculum resources, test items, and other educational tools marketed through the Internet. Curriculum messages delivered in this way tend to be fragmented and imprecise, and often misaligned.

CURRICULUM POLICY AND PRACTICE AT THE STATE LEVEL: A BATTLE FOR IDENTITY AND CONTROL

Adopting the Common Core Standards allows us to retain our standing as a state that holds all students to high academic expectations. These standards will spur academic achievement in the classroom. This decision also puts us right where we should be—at the table with other states to collaborate on innovative curricular and instructional strategies that will benefit students and educators for years to come.
—Massachusetts Education Commissioner Mitchell D. Chester, 2010

I firmly believe that states like Texas, working with local educators, employers, and citizens, are best suited to determine the curriculum standards for their students—not the federal government.
—Texas Governor Rick Perry, 2010

These two quotations represent contrasting ways that states have responded to the rollout of the CCSS in the United States. The quotation from the Commissioner of Massachusetts demonstrates his view of the CCSS as a collaborative project among states. In contrast, the Texas governor saw the standards as an intrusion of the federal government on educational policy, which is rightfully the jurisdiction of the state. In this section, we look more closely at the role that states and local communities play in determining the mathematics curriculum. We use different reactions to the CCSS initiatives to illustrate tensions that can emerge between local, state, and federal decision makers around curriculum policy. These local reactions highlight two major points of tension that have plagued the efforts to roll out the new standards. The first, a desire to maintain local control over curriculum and assessment, is a tension that is quintessentially American, but especially challenging in an increasingly interconnected society. The second tension, resistance to increased use of testing to measure student achievement and teacher and school effectiveness, is a more contemporary challenge faced by curriculum reformers.

Common State Standards: Actions and Reactions at the State Level

The CCSS initiative, which was originated by state governors, was recognized at the time as an unprecedented collaborative effort by state leaders to respond to shared educational challenges. The expectation was that once the standards were drafted, reviewed, and released, states would voluntarily adopt them, replacing a smorgasbord of state standards with a commonly shared set. Indeed, between 2010 and 2012, 45 states and the District of Columbia formally adopted the CCSSM. According to a provision of the CCSS initiative, adopting the standards meant state officials agreed that the core standards would constitute at least 85% of their state standards; the *15% rule* granted states the ability to add additional standards or make other modifications if they chose (Kendall, Ryan, Alpert, Richardson, & Schwols, 2012).

At the time of this writing, most of the states that originally adopted the CCSSM continue to use them as their official standards for mathematics; 18 states, along with the District of Columbia, use the CCSSM verbatim. Twenty states adopted the CCSSM with modifications. Five states chose not to adopt the CCSSM from the outset: Alaska, Minnesota, Nebraska, Texas, and Virginia (Certica Solutions, n.d.; see Table 6.3).

Among states that did not adopt the CCSSM, the professed reasons ranged from disagreements with the content of the standards to concerns about state control. The governor of Minnesota, for instance, claimed that the state's existing mathematics standards were sufficiently rigorous, and

TABLE 6.3 CCSSM Adoption Decisions by the 50 States and the District of Columbia as of 2016

Adopted Verbatim	Adopted With Modifications	Adopted Then Withdrew	Never Adopted
Connecticut	Alabama	Indiana	Alaska
Delaware	Arizona	Louisiana	Minnesota
District of Columbia	Arkansas	Michigan	Nebraska
Hawaii	California	Missouri	Texas
Idaho	Colorado	Oklahoma	Virginia
Illinois	Florida	South Carolina	
Kentucky	Georgia	Tennessee	
Maine	Iowa		
Mississippi	Kansas		
Nevada	Maryland		
New Hampshire	Massachusetts		
New Jersey	Montana		
Ohio	New Mexico		
Rhode Island	New York		
South Dakota	North Carolina		
Vermont	North Dakota		
Washington	Oregon		
Wisconsin	Pennsylvania		
Wyoming	Utah		
	West Virginia		

Source: Certica Solutions (n.d.).

adopting the CCSSM would "water down Minnesota's rigorous standards that require students to take algebra by eighth grade," jeopardizing the success that students in the state had demonstrated in mathematics (Pawlenty, 2010). In Texas, another state that rejected the CCSS, state officials emphasized a strong desire to maintain local control over educational policy. In a letter to the U.S. Secretary of Education, Texas governor Rick Perry cited the enormous expense of moving to a new set of standards, questioned the CCCS development process, and repeatedly asserted a desire to maintain state control: "In the interest of preserving our state sovereignty over matters concerning education and shielding our local schools from unwarranted federal intrusion into local district decision-making, Texas will not be submitting an application for RTTT funds" (Perry, 2010).

Reactions to the CCSS initiative by a number of states shifted over time. Although 45 states and the District of Columbia were initially quick to adopt the CCSSM, opposition began to fester in state offices of education, among local school districts, and within the general population. Between 2014 and 2016, seven states withdrew their adoption of the CCSSM, citing concerns about the federal government's influence in the initiative.

Testing Fatigue: Local Reactions to the CCSS

As described earlier, a key component of the CCSS initiative involved measuring state-level progress toward attaining the ambitious achievement goals set out in the standards. ED funded the development of assessments that would align with the new standards. Many stakeholders saw the standards and the assessments as two parts of an unwanted package. Within states that adopted the CCSSM, grassroots opposition to these new assessments fomented harsh opposition to the CCSS in general.

Major opposition came in response to the sheer amount of testing to which students were being subjected. As described earlier, the CCSS initiative arrived on the heels of the NCLB era, which had dramatically increased the use of assessments to measure educational achievement. States had also begun to tie teacher pay to student test scores, raising the stakes associated with these tests even higher. Many districts instituted additional practice and diagnostic tests throughout the year, to gauge progress and prepare students for these high-stakes tests. Educators, parents, and students began to decry the loss of instructional time and an inappropriate focus on test taking over learning.

A number of critics also questioned the quality of the new tests, which were developed rapidly and integrated new technology that had not previously been widely used for high-stakes tests. By some accounts, schools were being forced to purchase new computers in order to implement the assessments. Commentators also pointed to multiple flaws in the language and approach of many released items.

Because of these concerns, activists in many states joined the national "opt-out of testing" movement, described earlier. In 2015, when the new CCSS-aligned tests were first given, parents of over 650,000 students refused to allow their children to be tested (Fairtest.org, n.d.). In fact, by the 2015–2016 school year, only seven of the 26 states (including the District of Columbia) that had initially joined PARCC chose to fully use the tests, and 14 of the original 31 states that joined the SBAC opted to use that assessment (Duncan, 2010; Gewertz, 2016).

The Mathematics Curriculum as Determined by the States: A Delicate Balance

States have primary authority over educational policy in the United States, therefore we must look at states' actions and reactions to initiatives like the CCSSM to understand the official mathematics curriculum. First, it is important to recognize that the creation and adoption of the CCSS in the United States represents a landmark development in curriculum policy. Currently, 38 states and the District of Columbia have adopted the CCSSM as is or with modifications. Moreover, there is evidence that several states that withdrew from the CCSS initiative have adopted mathematics standards that resemble the Common Core. The CCSS represent unprecedented movement toward a coherent and consistent curriculum across the United States.

Second, the state and local reactions to outside imposition on curriculum and policy demonstrate the barriers to one of the aims of the CCSS initiatives, that states might combine resources and even leverage collective influence on the development of curriculum resources and assessments. A movement that began as a cross-state collaboration became viewed as a federal imposition, once the Department of Education stepped in with incentives and supports.

INSTRUCTIONAL RESOURCES: WHERE CURRICULUM POLICY MEETS CLASSROOM PRACTICE

[Textbooks] are intended as mediators between the intentions of the designers of curriculum policy and the teachers that provide instruction in classrooms.
—Valverde, Bianchi, Wolfe, Schmidt, Houang, 2002, p. 2

Instructional resources are a primary means by which expectations detailed in official curriculum documents are communicated to teachers. We now turn our focus to how instructional resources influence the mathematics curriculum in both the official and operational arenas in the United States.

We use the term *instructional resources* to refer to a wide array of materials available to teachers to design and enact instruction. These include print or digital textbooks and other comprehensive curriculum resources. Instructional resources also include supplemental materials, such as worksheets, instructional activities, and interactive tools, many of which can be found on the Internet.

In general, local school districts, or sometimes individual schools within these districts, have considerable autonomy in determining which instructional resources will be provided to teachers and how these resources should be used. In the case of 19 states, designated as *textbook-adoption* states, state policy provides an approved list of textbooks and regulates the timeline districts must follow when selecting new textbooks. We begin with an illustration of two districts that took substantially different paths after their states officially adopted the CCSSM.

Mathematics Curriculum Decisions in Two School Districts

Orange County is a large district located in Florida, a textbook-adoption state. After adopting CCSSM the state posted the list of approved curriculum materials. Orange County district administrators and teachers reviewed these programs and invited representatives from some of the publishing companies to present their offerings to district leaders. The county officials ultimately chose to purchase all their core K–12 mathematics curriculum materials from one of the major publishers. The full package included an option for digital or disposable student books as well as digital teacher editions. District officials then created sequencing and pacing guides, which laid out the content from Florida's mathematics standards (composed of the CCSSM and some additional standards) and designated specific standards teachers should cover in each quarter. The materials purchased from the major publisher were included as possible resources to use, but teachers were not mandated to use the new materials for their lessons if they preferred another resource (Gewertz, 2014).

A different approach was taken by the Long Beach Unified School District, located in California, which is another textbook-adoption state.[7] In 2014, the California Department of Education had not yet released its approved list of curriculum materials aligned to California's version of the CCSSM. So the district enlisted a team of teachers, content specialists, and instructional coaches to repurpose the curriculum materials they had purchased prior to the CCSSM to align to the new standards. They resequenced and omitted units in the old curriculum materials to match the new content and added or revised lessons to foster classroom discourse and emphasize problem solving. After determining how long the lessons and units should take, the team created pacing guidelines to ensure that teachers would cover the necessary material over the course of the year. Teachers in Long Beach were required to use the customized set of instructional resources that resulted from this revisioning process (Gewertz, 2014).

Determining the Designated Curriculum

The two district profiles illustrate different approaches to selecting and regulating the instructional resources available to teachers. These decisions are part of what Remillard and Heck (2014) refer to as the designated curriculum, or the means by which districts or schools provide guidance to teachers about how the standards should be addressed in the classroom. Instructional resources are incorporated into the designated curriculum, sometimes in the form of a single textbook, or as "a host of assembled packages of materials, instructional resources, and structuring guidelines" (p. 710). The level of detail provided by the district or school in the designated curriculum varies considerably, with some requiring teachers to follow a pacing guide that delineates which lesson should be covered on any given day and, in some cases, specific instructional materials that teachers are expected to use for each lesson.

What Do Instructional Resources Look Like, and Who Creates Them?

Since the NCTM *Standards* reforms in the 1990s and the rise of digital technologies in the early 21st century, the mathematics instructional materials market has expanded and diversified, both in terms of the types of offerings and who develops them. The CCSSM were rolled out during a transition from print to digital materials, and at a time when curriculum development extended beyond the auspices of mainstream publishers, to include a variety of smaller organizations, including teachers.

Mainstream Publishers and Conventional Curriculum Materials

In the United States, commercial publishers have long created the textbooks used in a majority of classrooms. These large companies were often the only publishers that could compete effectively in a marketplace where each state followed separate content standards and, in many cases, their own criteria for acceptable instructional materials (Seeley, 2003). Prior to the publication of the NCTM *Standards,* most commercial curriculum materials in the United States were fairly uniform in their appearance and instructional style; lessons began with examples of a specific skill, progressed through a generous set of practice problems, and usually ended with a selection of word problems (Eliot, 1990; Stein, Remillard, & Smith, 2007).

Innovative Curriculum Materials

Throughout the 20th century, there were instances in which U.S. mathematicians and mathematics educators created curriculum materials that varied from those produced commercially. These materials typically failed to gain popularity due to the complexity of the textbook adoption process across the many states in the United States (Seeley, 2003).

By contrast, the NSF-funded curriculum development project teams that developed materials to align to the NCTM *Standards* (1989) represented a turning point. In order to be awarded funding, the academic centers proposing to create new materials were required to partner with commercial publishers for the production of the print materials. Most of the academic centers partnered with smaller publishers to produce materials that *emphasize* problem solving and reasoning, rather that practicing procedures. Over time, smaller publishers were acquired by large companies. By the time the CCSSM were published, each of the big three publishers, Pearson, McGraw Hill Education, and Houghton Mifflin Harcourt, sold both conventional and *Standards*-based materials.

Common Core Aligned Materials

In response to the CCSSM, commercial publishers worked to update their conventional textbooks and have continued to work with the original authors to update their *Standards*-based curriculum materials to align to the CCSSM. Many of the features of the *Standards*-based materials, like the inclusion of cognitively demanding tasks and an emphasis on multiple representations and strategies, were adopted by the mainstream publishers to align with CCSSM. Questions remained about whether instructional materials that claimed alignment to the CCSSM actually fulfilled the characteristics identified in the standards. Polikoff (2015), for example, analyzed three "Common Core aligned" fourth grade math textbooks published in 2012 and found that, although they generally covered the content of the standards, the texts "systematically overemphasize procedures and memorization and underemphasize more conceptual skills relative to their emphasis in the standards" (p. 1185).

Digital Materials

Increasingly, publishers are transitioning to digital offerings. In 2016, most offer print materials, along with digital components, including online access to pdf replicas of print textbooks, videos, and interactive student activities. The movement to digital materials has been slow, in part because many school districts lack the technological infrastructure to support fully digital programs. A few curriculum designers have begun to develop instructional materials that have been described as "deeply digital" (Edson, 2014). More than simply providing digital copies of printed material along with occasional video and interactive applets, these materials leverage technology to provide new instructional interactions in classrooms, including dynamic instructional scaffolding that provides differentiated experiences for students. Digital assessments provide students with immediate feedback, and in some cases, adaptive computer tutors. Data can be fed back to students, parents, teachers, and lesson developers to inform future instructional decisions and materials development.

The Rise of Internet Publishing

Access to the Internet has allowed organizations not traditionally part of the publishing market to disseminate new instructional resources. Teachers Pay Teachers, for example, emerged to provide an outlet for teachers to publish and sell the instructional materials they create. A company called LearnZillion began to provide a full digital curriculum at no charge to districts or teachers. A number of states created instructional resources and made them available on the web. Most notably, the state of New York, using Race to the Top funding, partnered with external agencies to develop a comprehensive set of curriculum materials aligned with the CCSSM, then called Eureka Math (Engageny, n.d.). These instructional materials were published on the Internet and are freely downloaded by teachers throughout the country.

Supplemental Resources

The mathematics curriculum materials market in the United States also contains a plethora of supplemental materials, intended for remediation, test preparation, or instructional alternatives. Some materials address only a subset of the mathematical content and are designed to enhance a comprehensive curriculum resource. Other resources have been developed and distributed through websites, to support and guide teachers to understand the CCSSM and the related mathematical concepts more deeply. Two of the CCSSM authors, for example, created documents, videos, and other teaching tools to support the use of the Common Core.[8]

Decreased Reliance on a Single Comprehensive Curriculum

The diversification and expansion of the instructional materials industry, described above, has led researchers to explore what resources districts, schools, and teachers are using following publication of the CCSSM. This research is limited, given that each district in the United States is responsible for determining its own designated curriculum. Two recent studies, however, suggest movement away from a single comprehensive curriculum resource in favor of a variety of purchased and open-access resources.

A 2013 national survey of middle school teachers indicated that over two thirds were using textbooks adopted before the CCSSM were created and that, consequently, over 60% reported supplementing their primary curriculum using materials from the Internet (Davis, Chopin, Roth McDuffie, & Drake, 2013). Another 2013–2014 survey of 257 elementary school teachers and teacher leaders in a state using the CCSSM for the second year found that only about half of respondents reported using resources advertised as comprehensive curricula as their primary resource and, of those respondents who used the comprehensive curricula, only half reported using that curriculum for as much as 80% of their instruction (Polly, 2016). These results indicate that as the CCSSM were adopted, many teachers implemented what Polly calls "curriculum by compilation," or a patchwork of supplemental instructional materials sourced from a variety of for-profit and nonprofit entities.

The trend toward curriculum by compilation appears to be influenced by several factors. First, the textbooks that teachers had access to when the CCSSM were first adopted were not aligned with the new standards, forcing many districts to develop new resources or adapt their old materials. Second, the increased use of the Internet to disseminate and market curriculum resources has provided districts and teachers easy access to a wide array of supplemental materials. Third, following the U.S. financial crisis of 2008, many districts encountered the reduction or complete elimination of funds for textbooks and district-level mathematics specialists; free resources from the Internet were an increasingly attractive option in this resource-scarce environment.

Trends in the Development and Use of Instructional Resources

The instructional resources found in U.S. classrooms have always been diverse, due to the fact that decisions about instructional materials are made at the local level. But the developments described in this section indicate that, in the CCSSM era, options have become even more varied. Districts can choose between innovative materials or more traditional approaches, available in print and digital forms. Furthermore, the ease of Internet publishing has increased access to materials from many different types of sources, including but not limited to established publishers, smaller for-profit and nonprofit organizations, state and district departments of education, and teachers themselves.

The trend toward sourcing the Internet for instructional materials and interactive tools increasingly positions the teacher as determiner of which resources to use to teach any given lesson (Dorsey, 2016), placing significant demand on teachers. The vast collection of CCSSM-aligned instructional material varies considerably in terms of pedagogical orientation, mathematical accuracy, actual alignment to the CCSSM, and to current research-based recommendations from the mathematics education research community. As they sift through the vast assortment of offerings, teachers must possess a great deal of curricular expertise to synthesize a cohesive curriculum composed of high quality instructional experiences. If districts and teachers continue to create these patchwork curriculum plans, it remains to be seen whether students in the United States will experience the coherent curriculum that the CCSSM were designed to achieve.

TEACHER AS ENACTOR OF CURRICULUM

The mediating role of textbooks is complex because teachers and students are not passive in creating educational opportunities. Each has a unique impact on what transpires in classrooms.
—Valverde et al., 2002, p. 10

We now shift our discussion of the mathematics curriculum in the United States from the official to the operational curriculum (Remillard & Heck, 2014). At this stage, individual teachers and teacher groups play a critical role in shaping the curriculum. The enactment process begins with teachers designing an intended or planned curriculum using available resources. It also includes the curriculum that is enacted in the classroom with students, in a joint process of design.

Images of Teachers Designing Curriculum

In this section, we focus primarily on consequential aspects of what is known about the variety of ways teachers interact with curriculum resources to design lessons. The teachers in the following vignettes are fictitious in that they do not represent individual teachers. Rather, their stories are aggregated from research findings and first-hand experience of both authors in order to illustrate a wide range of teachers' experiences designing mathematics curriculum.

Curriculum by Compilation

Carmen teaches third grade in a large city school district. As in many urban districts, the budget for instructional materials is minimal. Rather than purchasing new CCSSM-aligned textbooks, district leaders provided teachers a pacing guide and recommended they access free resources from the Internet. When planning a lesson focused on CCSSM standard 3.NF.A.2: "Understand a fraction as a number on the number line; represent fractions on a number line diagram" (NGACBP & CCSSO, 2010), Carmen first reviews the two lessons recommended by her district pacing guide. One lesson is from Eureka Math, the comprehensive curriculum program commissioned by the State of New York, and another is offered by LearnZillion, an open, cloud-based curriculum resource.

Carmen chooses to use a *conceptual development* activity involving folding paper strips from an Eureka Math lesson.[9] She then selects a video from LearnZillion that shows a transition from fraction strips to the number line.[10] To provide an opportunity for her students to practice, she selects a set of in-class student exercises from the Eureka Math lesson; she also chooses to assign half of the Eureka Math homework assignment. She then finds a multiplication practice sheet from the Internet to assign, because she worries that her students are not yet fluent with their multiplication facts. In order to select a warm-up activity, Carmen logs onto her account on the test preparation engine purchased by her district. In the Grade 3 section she navigates to two standards covered in the previous unit and downloads a file, which she plans to project on the board at the beginning of class.

Diving Into Digital

Dianne teaches fifth grade in a well-funded suburban district, which has purchased a comprehensive digital resource, developed by a major publisher. Each of her students has a tablet, which they use to access the lessons and related activities.

Dianne's next lesson focuses on CCSSM 5.NF.A.1 "Add and subtract fractions with unlike denominators (including mixed numbers) by replacing given fractions with equivalent fractions in such a way as to produce an equivalent sum or difference of fractions with like denominators" (NGACBP & CCSSO, 2010). Because the curriculum resource is written to align with the CCSSM, she does not feel the need to consult her state's standards, which follow the CCSSM closely. Instead, she navigates directly to the lesson using the digital teacher's guide, knowing that the lesson will contain the text of the standards it addresses. Dianne uses all the lessons in each chapter, never changing the sequence. The district has provided a pacing guide. Dianne tries to follow it, slowing down only if she feels a majority of the class is having trouble understanding the material.

She begins the planning process by reviewing the aggregated results from a recent quiz in the analytic tool of the program and identifies the two questions with which students had the most trouble. She creates multiple choice questions similar to these items, to provide students practice with the sorts of questions they might encounter on the test at the end of the year. She also chooses an introductory video that explains how carpenters sometimes add fractions when building furniture. She next selects the main task she will present to the class to introduce the concept of adding fractions, a problem set in a woodworking shop. Then, she reviews how to launch and use the virtual manipulatives, which allow students to create fractions by equipartitioning rectangular area models. Finally, Dianne looks at the "*student dashboard*," which records students' progress on the individualized virtual tutoring system. She makes note of those students who have not been active lately, and makes plans to follow up.

Adapting an Aging Standards-Based Resource

John teaches middle school in a small city that purchased a *Standards*-based curriculum resource in 2008. Because the district has not yet purchased the new CCSSM-aligned edition, John teaches from the old version, omitting lessons and adding others to comply with his state's standards, which are based on the CCSSM. He draws most of his added lessons from a "CCSSM aligned supplement," developed by the curriculum publisher to bridge the period between the release of the CCSSM and publication of the comprehensive CCSSM edition of the curriculum materials. He looks forward to receiving the revised edition, which his district intends to purchase for the upcoming school year, the scheduled year for the next mathematics curriculum program adoption.

John and the other seventh-grade math teachers meet with the district math coach to plan the lessons for the following week. They identify the first standard they will address: CCSSM standard 7.RP.A.1 "Compute unit rates associated with ratios of fractions, including ratios of lengths, areas, and other quantities measured in like or different units. *For example, if a person walks ½ mile in each ¼ hour, compute the unit rate as the complex fraction ½ / ¼ miles per hour, equivalently 2 miles per hour*" (NGACBP & CCSSO, 2010, p. 48). The teachers then read through a lesson addressing this standard in their students' textbook. When skimming through the one teacher's guide that the three teachers share, his colleague notices a section on common student responses to the activity they planned. They review these together and plan their own reactions for each type of student response. They select an extension to the task they can assign to students they foresee will finish ahead of the others. Then they read through the sample classroom dialogue provided in the teacher guide, talk through the questions they want to use during the discussion, and plan how they want the whiteboard to look when they are finished.

Curriculum Making in the Classroom

Carmen, Dianne, and John and his colleagues provide three examples of different approaches to using existing resources to design curriculum. As we discussed at the beginning of the section, all three teachers illustrate the role the teacher plays in curriculum design. This role is most likely universal, since, around the world, teachers find themselves interpreting components of the official curriculum documents and resources and making design decisions that suit their local context. The three vignettes also illustrate particular characteristics of this process in the contemporary U.S. curriculum system: variation, compilation, and the influence of tests.

Variation is the Norm

School districts determine the designated curriculum, which specifies the guiding frameworks and resources available to teachers and the degree of flexibility they have within these structures. The type and nature of resources can vary substantially from district to district. How the standards are communicated to teachers differs, as does the quality of instructional materials they are provided. As a result, teachers encounter different representations of the standards and have uneven opportunities to learn about them. They also have differing opportunities to learn about the CCSSM through professional development. Another form of variation represented in the vignettes is how teachers interact with resources to design their curriculum. These interactions are influenced by teachers' knowledge and beliefs, professional identity, and their stance toward curriculum materials (Remillard, 2005). All three of the vignettes show teachers using resources and guides for teachers. Not all teachers do this. Some teachers in the United States use the problems in the student textbook to design lessons and never consult the guidance that is contained in the guide for teachers (Remillard & Bryans, 2004; Reinke, 2015).

An important issue, highlighted by the variation across the vignettes, is the reality that not all curriculum resources are created equally. The three examples contrast a compilation of freely available materials with a fully digital curriculum program—developed to align with the CCSSM and complete with analytic tools that allow teachers to monitor student progress, and a not fully updated program—requiring regular supplementation. Because funding for education in the United States is determined primarily by local taxes, curriculum resource purchases are often determined by the wealth of the community.

Movement Toward Compilation in a Connected World

It is increasingly common in the United States for teachers, often directed by school districts, to build their intended curriculum through compilation. Similar to Carmen, some districts point teachers to many resources from which they are expected to draw and assemble a program of study. Even in districts that provide a textbook or other primary curricular resource, many teachers supplement lessons with tasks and drills that provide additional practice on computational skills or other items likely to appear on the state tests. The compilation approach is facilitated by the online marketplace of resources, where teachers can easily browse for and obtain a range of instructional resources beyond those adopted by their district.

The Test as the Primary Messenger of the Standards

Although teachers appear to be accessing information about their state standards from differing sources, it is evident that many teachers use a high-stakes test as the primary means of information about what is expected of students and, consequently, what they are expected to teach. This tendency can be traced to the accountability movement, which dominated the early 2000s (NCLB Act, 2001), and the decision to use assessments as a key driver of the CCSS initiative. *Test prep* tasks were routinely integrated into regular instruction. In many states, new tests, based on the CCSS, have replaced the old tests, but the pressures surrounding them have remained the same.

The Teacher and the Design of the Operational Curriculum

As the quotation opening this section suggests, teachers (and students) uniquely impact the curriculum that transpires in the classroom. From this perspective, teachers should be understood as designers of local curriculum. This design work, however, occurs within a context of state and local policy and is highly influenced by state standards and assessments, the instructional resources made available to them, and the degree of flexibility granted. In the current U.S. context, we are seeing increased expectations that teachers will play a compiling or curating role in curriculum program design. This expectation appears to be bolstered by Internet connectivity and the free market, which have increased teachers' access to a range of resources. It also represents a notable shift from practices of the recent past, when a majority of teachers assigned work from a comprehensive textbook on a daily basis (Braswell et al., 2001) and did not have easy access to alternative resources. A related dimension of the curriculum compiling trend is that teachers are expected to consult policy documents, such as state standards and released test items, to guide their design decisions. Consequently, as the role of specific curriculum programs decreases, the demands on teachers are intensified. It then follows that the quality of the enacted mathematics curriculum is highly dependent on teacher capacity, access to resources, and their vision about good teaching.

THE U.S. MATHEMATICS CURRICULUM: OPPORTUNITIES AND CHALLENGES

Our examination of the official and operational mathematics curriculum in the United States reveals long-standing characteristics inherent to education in the United States, including a tendency toward state and local control and strong market influences. These characteristics are unlikely to change and will continue to contribute to variation across states, districts, and individual classrooms. The CCSS initiative represents an unprecedented move toward greater coherence across the majority of U.S. states. Our look at how this initiative is playing out at different levels of the U.S. educational system reveals opportunities and challenges for curricular coherence.

The most promising opportunity may be in the fact that the United States is closer to a common set of mathematics learning goals than ever before. The widespread, state-by-state adoption of a core set of standards in mathematics is significant. Equally promising, many would argue, is the intentional design of these standards around focus, coherence, and a complex view of rigor, fashioning them after the curriculum structures of top-performing countries (McCallum, 2015). With this foundation in place, those developing instructional resources have the opportunity to focus on fewer sets of standards, rather than one for every state. The fact that an increasing number of instructional resources are distributed to teachers through the Internet increases the chance that curriculum resources will be shared by teachers across states, even in states that have not adopted the CCSSM.

Alongside these opportunities, we see several challenges, or threats, to curricular coherence. These challenges are largely related to the approach used to implement the CCSS movement. For instance, the use of the RTTT competition to entice states to adopt the standards and the federal funding of assessment programs represent an unparalleled level of involvement in state curriculum matters by the federal government. Although these federal actions undoubtedly encouraged states to adopt the CCSS, the involvement of the federal government may have come at a cost. The CCSS initiative began as a partnership among state education leaders; once the federal government stepped in, state leaders concerned with local control over education began to retreat. Further, the decision to emphasize accountability, leveraged through testing, has also had a damaging effect on the reception of the CCSS. Across the country, the new standards and the new tests have become inextricably linked and public resistance to testing has also been directed to the CCSS initiative in general. This resistance continues to threaten the prospects of the CCSSM.

Another challenge to the potential coherence sought by the CCSSM is the decision to leave resource development and specific support for teachers to the private marketplace, which has produced a plethora of varied and fragmented resources. Consequently, teachers find themselves navigating an uneven landscape, compiling a curriculum with insufficient support. It may be the lack of support for the curriculum work undertaken by teachers that provides the greatest threat to curricular coherence.

NOTES

1. For detail on how the U.S. school system is structured, see Dossey, McCrone, and Halvorsen (2016).
2. Two sets of standards were developed, one for English Language Arts and one for Mathematics. The acronym CCSSM refers to the standards for mathematics, which is the focus of this chapter. We use CCSS when referring to both sets of Standards and CCSS initiative to refer to the full initiative laid out in the *Benchmarking for Success* report (National Governors Association, the Council of Chief State School Officers, & Achieve, Inc., 2008).
3. Summary reports from these reviews are available online (see http://www.corestandards.org).
4. Both Publishers' Criteria for the Common Core for Mathematics can be found on the Common Core website (http://www.corestandards.org/other-resources/).
5. Private schools, which are supported by tuition fees and other revenues rather than state and local tax funds, are fully autonomous of state curriculum policy.
6. See Schoenfeld (2004) for a detailed account of the math wars.
7. California is a textbook adoption state at the elementary level (Grades K–8). The state does not approve textbooks for the secondary level.
8. William McCallum founded Illustrative Mathematics, the non-profit organization responsible for illustrativemathematics.org and Jason Zimba co-founded Student Achievement Partners, the non-profit organization responsible for achievethecore.org.
9. https://www.engageny.org/file/35366/download/math-g3-m5-topic-d-lesson-14.pdf?token=fcv_5xrG
10. https://learnzillion.com/lesson_plans/3464-3-identify-and-locate-fractions-fp

REFERENCES

Apple, M. W. (1992). Do the standards go far enough? Power, policy, and practice in mathematics education. *Journal for Research in Mathematics Education, 23*(5), 412–431.

Ball, D. L., & Cohen, D. K. (1996). Reform by the book: What is: Or might be: The role of curriculum materials in teacher learning and instructional reform? *Educational Researcher, 25*(9), 6–14.

Braswell, J. S., Lutkus, A. D., Grigg, W. S., Santapau, S. L., Tay-Lim, B., & Johnson, M. (2001). *The nation's report card: Mathematics 2000.* NCES 2001-517. Washington, DC: U.S. Department of Education, Office of Educational Research and Improvement, National Center for Education Statistics.

Certica Solutions. (n.d.). *Common Core State Standards Adoption Map.* Retrieved from http://statestandards.staging.wpengine.com/common-core-state-adoption-map/

Chicago Teachers Union. (2014, July). *Arguments against the Common Core.* Retrieved from http://www.ctunet.com/quest-center/research/text/CTU-Common-Core-Position-Paper.pdf

Chingos, M. M., & Whitehurst G. J. (2012). *Choosing blindly: Instructional materials, teacher effectiveness, and the Common Core.* Providence, RI: Policy Report Brown Center on Educational Policy at Brookings.

Common Core State Standards Initiative. (n.d.a). *Read the Standards.* Retrieved from http://www.corestandards.org/read-the-standards/

Common Core State Standards Initiative. (n.d.b). *Standards for Mathematical Practice.* Retrieved from http://www.corestandards.org/Math/Practice/

Council of Chief State School Officers [CCSSO]. (1995). *State curriculum frameworks in mathematics and science: How are they changing across the states?* Washington, DC: Author.

Davis, J., Choppin, J., McDuffie, A. R., & Drake, C. (2013). *Common core state standards for mathematics: Middle school mathematics teachers' perceptions.* Rochester, NY: The Warner Center for Professional Development and Education Reform: University of Rochester. Retrieved from https://www.warner.rochester.edu/files/warnercenter/docs/commoncoremathreport.pdf

Dorsey, C. (2016). Deeply digital STEM learning. Keeping an eye on the teacher in the digital curriculum race. In M. Bates & Z. Usiskin (Eds.), *Digital curricula in school mathematics* (pp. 285-296). Charlotte, NC: Information Age.

Dossey, J., McCrone, S., & Halvorsen, K. (2016). *Mathematics education in the United States 2016: A capsule summary fact book.* Reston, VA: National Council of Teachers of Mathematics. Retrieved from http://www.nctm.org/uploadedFiles/About/MathEdInUS2016.pdf

Duncan, A. (2010, September 2). *Beyond the bubble tests: The next generation of assessments—Secretary Arne Duncan's Remarks to State Leaders at Achieve's American Diploma Project Leadership Team Meeting.* Retrieved from http://www.ed.gov/news/speeches/beyond-bubble-tests-next-generation-assessments-secretary-arne-duncans-remarks-state-leaders-achieves-american-diploma-project-leadership-team-meeting

Edson, A. J. (2014). *A deeply digital instructional unit on binomial distributions and statistical inference: A design experiment.* Doctoral Dissertation, Western Michigan University. Retrieved from http://scholarworks.wmich.edu/dissertations/275

Eliot, D. L. (1990). Textbooks and the curriculum in the postwar era: 1950-1980. In D. L. Elliot & A. Woodward (Eds.), *Textbooks and schooling in the United States* (pp. 42–55). Chicago, IL: University of Chicago Press.

engageny. (n.d.). *Frequently asked questions.* Retrieved from https://www.engageny.org/frequently-asked-questions

Franklin, C. (2013). Common Core State Standards and the future of teacher preparation in statistics. *The Mathematics Educator, 22*(2), 3–10.

Gewertz, C. (2014, April 21). Two districts, two tacks on curriculum. *Education Week.* Retrieved from http://www.edweek.org/ew/articles/2014/04/23/29cc-curriculum.h33.html

Gewertz, C. (2016, March 24). State testing: An interactive breakdown of 2015-16 plans. *Education Week.* Retrieved from http://www.edweek.org/ew/section/multimedia/state-testing-an-interactive-breakdown-of-2015-16.html

Kendall, J., Ryan, S., Alpert, A., Richardson, A., & Schwols, A. (2012, March). *State adoption of the Common Core State Standards: The 15 percent rule.* Mid-continent Research for Education and Learning (McREL). Retrieved from http://files.eric.ed.gov/fulltext/ED544664.pdf

Martin, D. B. (2003). Hidden assumptions and unaddressed questions in mathematics for all rhetoric. *The Mathematics Educator, 13*(2), 7–21.

Massachusetts Department of Elementary and Secondary Education. (2010). Minutes of the Special Meeting of the Massachusetts Board of Elementary and Secondary Education. July 21, 2010. Retrieved from http://www.doe.mass.edu/boe/minutes/10/0721reg.pdf

McCallum, W. (2015). The Common Core State Standards in Mathematics. In S. J. Cho (Ed.), *Selected regular lectures from the 12th International Congress on Mathematical Education* (pp. 547–561). Switzerland: Springer International.

National Center for Educational Statistics. (n.d.). *Characteristics of the 100 largest public elementary and secondary school districts in the United States: 2000–2001.* Retrieved from https://nces.ed.gov/pubs2002/100_largest/table_app_a_1.asp

National Council of Teachers of Mathematics [NCTM]. (1989). *Curriculum and evaluation standards for school mathematics.* Reston, VA: Author.

National Council of Teachers of Mathematics [NCTM]. (2000). *Principles and standards for school Mathematics.* Reston, VA: Author.

National Governors Association Center for Best Practices [NGACBP], & Council of Chief State School Officers [CCSSO]. (2010). *Common Core State Standards for Mathematics.* National Governors Association Center for Best Practices, Council of Chief State School Officers, Washington D.C.

National Governors Association [NGA], Council of Chief State School Officers [CCSSO], & Achieve, Inc. (2008). *Benchmarking for success: Ensuring U.S. students receive a world-class education.* A report by the NGA, CCSSO, and Achieve. Washington, D.C.

National Governors Association [NGA], Council of Chief State School Officers [CCSSO], Achieve, Council of the Great City Schools, National Association of State Boards of Education. (2013a). *K–8 publishers' criteria—Common Core*

State Standards Initiative. Retrieved from http://www.corestandards.org/wp-content/uploads/Math_Publishers_Criteria_K-8_Spring_2013_FINAL1.pdf

National Governors Association [NGA], Council of Chief State School Officers [CCSSO], Achieve, Council of the Great City Schools, National Association of State Boards of Education. (2013b). *High school publishers' criteria–Common Core State Standards Initiative*. Retrieved from http://www.corestandards.org/assets/Math_Publishers_Criteria_HS_Spring%202013_FINAL.pdf

National Research Council [NRC]. (2001). *Adding it up: Helping children learn mathematics.* J. Kilpatrick, J. Swafford, and B. Findell (Eds.). Mathematics Learning Study Committee, Center for Education, Division of Behavioral and Social Sciences and Education. Washington, DC: National Academy Press.

No Child Left Behind [NCLB] Act. (2001). Public Law 107–110. 107th Congress. Retrieved from http://www2.ed.gov/policy/elsec/leg/esea02/107-110.pdf

Pawlenty, T. (2010, March 10). A Statement from the Governor [Press Release]. Retrieved from http://www.leg.state.mn.us/docs/2010/other/101583/www.governor.state.mn.us/mediacenter/pressreleases/PROD009899.html

Perry, R. (2010, January). Open Letter to the Honorable Arne Duncan, Secretary of Education. Retrieved from http://www.lrl.state.tx.us/scanned/govdocs/Rick%20Perry/2010/pressrelease011310.pdf

Polikoff, M. S. (2015). How well aligned are textbooks to the Common Core Standards in Mathematics? *American Educational Research Journal, 52*(6), 1185–1211.

Polly, D. (2016). Elementary school teachers' uses of mathematics curricular resources. *Journal of Curriculum Studies*. DOI: 10.1080/00220272.2016.1154608

Reinke, L. (2015). *Characterizing the role of contextualized problems in a written and enacted algebra unit.* Unpublished doctoral dissertation, University of Pennsylvania, Philadelphia, PA.

Remillard, J. T. (2005). Examining key concepts in research on teachers' use of mathematics curricula. *Review of Educational Research, 75*(2), 211–246.

Remillard, J. T., & Bryans, M. B. (2004). Teachers' orientations toward mathematics curriculum materials: Implications for teacher learning. *Journal for Research in Mathematics Education, 35*, 352–388.

Remillard, J. T., & Heck, D. (2014). Conceptualizing the curriculum enactment process in mathematics education. *ZDM: The International Journal on Mathematics Education, 46*(5), 705–718.

Remillard, J. T., & Taton, J. (2015). Rewriting myths about curriculum materials and teaching to new standards. In J. A. Supovitz & J. Spillane (Eds.), *Challenging standards: Navigating conflict and building capacity in the era of the common core* (pp. 49–58). Lahnam, MD: Rowan & Littlefield.

Reys, B. J. (Ed.). (2006). *The intended mathematics curriculum as represented in state-level curriculum standards: Consensus or confusion?* Charlotte, NC: Information Age.

Schmidt, W., Houang, R., & Cogan, L. (2002, Summer). A coherent curriculum: The case of mathematics. *American Educator*, 1–17.

Schmidt, W. H., Wang, H. C., & McKnight, C. C. (2005). Curriculum coherence: An examination of US mathematics and science content standards from an international perspective. *Journal of Curriculum Studies, 37*(5), 525–559.

Schoenfeld, A. H. (2004). The math wars. *Educational Policy, 18*(1), 253–286.

Seeley, C. (2003). Mathematics textbook adoption in the United States. In G. Stanic & J. Kilpatrick (Eds.), *A history of mathematics education* (Vol. 2, pp. 957–988). Reston, VA: National Council of Teachers of Mathematics.

Senk, S. L., & Thompson, D. R. (2003). *Standards-based school mathematics curricula: What are they? What do students learn?* Mahwah, NJ: Lawrence Erlbaum Associates.

Stein, M. K., Remillard, J. T., & Smith, M. S. (2007). How curriculum influences student learning. In F. Lester (Ed.), *Second handbook of research on mathematics teaching and learning* (pp. 319–369). Charlotte, NC: Information Age.

U.S. Department of Education. (n.d.). *States' applications for phase 2*. Retrieved from https://www2.ed.gov/programs/racetothetop/phase2-applications/index.html

Valverde, G. A., Bianchi, L. J., Wolfe, R., Schmidt, W. H., & Houang, R. T. (2002). *According to the book: Using TIMSS to investigate the translation of policy into practice through the world of textbooks*. Dordrecht, the Netherlands: Kluwer.

Wolfmeyer, M. (2014). *Math education for America?: Policy networks, big business, and pedagogy wars*. New York, NY: Routledge.

CHAPTER 7

A SOUTH AFRICAN PERSPECTIVE ON THE MATHEMATICS CURRICULUM

Towards a Transcending Metacognitive Ideology

Divan Jagals and Marthie van der Walt
North-West University

In this chapter, we offer an overview of South Africa's mathematics education landscape, including a discussion of the evolution of the mathematics curriculum. We focus on curriculum change and theoretical developments as constructs of curriculum transformation in this country. We consider the transcending issues of language, curriculum for the 21st century, and the promotion of metacognitive thinking embedded (or not) in the mathematics curriculum. Relevant research on the mathematics curriculum provides a basis for the South African perspective and recommendations for more emphasis on depth of mathematical learning experiences, rather than on a broad scope of mathematics topics. However, we argue that there is a transcending metacognitive ideology that is missing from the South African mathematics curriculum as the curriculum is content loaded and does not offer the depth of knowledge and understanding needed to prepare learners adequately for their future, particularly as it overlooks many aspects associated with metacognitive thinking. Consistent with the country's historical perspective, we advocate for a shift in the way curriculum developers and implementers think about the curriculum and elaborate on current work in the field for moving this thinking forward. We combine theory with practice to respond to the transcending metacognitive ideology embedded in South Africa's mathematics curriculum, a construct that is often overlooked in curriculum development.

THE CHARACTER OF SOUTH AFRICA'S EDUCATION IDEOLOGIES

The term *ideology* refers to theoretical and practically established systems that shape the assumptions about a society's social norms and relations. This is usually done to achieve, what is believed to be, a utopian worldview on the structure of society. In mathematics education, ideologies take the form of beliefs about mathematics as a school subject and about one's abilities to do mathematics. Typical examples of ideologies in education include recent debates arguing for or against outcomes or objectives, direct instruction or cooperative learning, and behaviorist or constructivist approaches. Naturally, these debates revolve around two central issues: *how one should teach mathematics* and *how one should learn mathematics*. Each view attempts to dominate the other, until only the seeming *correct* solution, or *way* of teaching, remains standing. Education ideologies, therefore, serve as theoretical truths or educational beliefs that underlie the quality of schooling (Samuelsson & Samuelsson, 2016). Ernest (1991, 2010) identified five types of education ideologies: the industrial trainer, the technological pragmatist, the old humanist, the progressive educator, and the public educator. We propose metacognitive ideology as a sixth type of ideology.

Each of Ernest's (2010) ideologies has the potential for academic success in mathematics, yet each has limitations. For example, *industrial trainers* usually see mathematics as a pure, formula-based, rule-driven tool to understand our world; there is a risk of a learner not knowing what he or she knows, but anticipates what the teacher wants him or her to know. *Technological pragmatists* consider mathematical knowledge to have practical use acquired through experience; this view prescribes content (or theory) and practice together, but does not focus on contextually relevant practical problems. The *old humanist* view regards mathematics as a structure of pure knowledge often learned through teachers' direct examples of rote-memorized skills and steps, yet leaves little room for thinking about one's own thinking when merely imitating the steps of others. The fourth ideology is the *progressive educator*, who facilitates exploring mathematical concepts through play and active involvement in the discovery process. Although this view fits with a hands-on approach to connect theory with practice, it requires abstraction and mathematizing capabilities for critical thinking and evaluation. Space and time for the development of such thinking are not necessarily possible in a curriculum that is too full. Samuelsson and Samuelsson (2016) noted Ernest's (2010) fifth ideology as the *public educator*, in which the learning of mathematics is socially mediated through a series of dialogues, questioning, and pedagogies. With this view, students need to responsibly regulate their own and others' learning, a skill that requires awareness of and modeling by teachers of their ways of thinking. The product of these diverse views shows that mathematics can be considered as a socially responsive discipline.

This social understanding of mathematics addresses what Schoenfeld (2013) identifies as necessary skills to analyze one's own problem solving attempts. He mentions four skills for this awareness to analyze problem solving actions/progress: the individual's knowledge, the use of strategies, the monitoring of performance, and the belief system of the individual's level of success. These skills link closely with what Flavell (1979) refers to as *metacognition*, or being aware of the nature of one's knowing. Referring back to the ideologies of Ernest, an *industrial trainer* does not promote this metacognitive thinking, a *technological pragmatist* does not necessitate a belief in one's own success, nor does the *old humanist* view leave a sense of wonder about one's own knowing. We are left with either a *progressive* or *public educator*, who has the potential to create opportunities for play and discovery, and debate the needed knowledge and skills. Both these ideologies seem promising for what the current curriculum in South Africa offers, yet the element of thinking about one's thinking in the curriculum is only indirectly implied or absent in the curriculum documents.

The point we are making is this: education ideologies are contextualized by both the mathematics content and the social context of our curriculum. Different content, contexts, and approaches can leave some ideological absolutes invalid. To offer our view of *metacognition* as a transcending ideology amidst these conditions, we first explore the context of South Africa's mathematics education landscape.

THE CONTEXT OF SOUTH AFRICA'S MATHEMATICS EDUCATION LANDSCAPE

We consider the incompleteness of ideological absolutes previously described in the South African education system with awareness of a contemporary issue: the relationship between race and socioeconomic success.

One side of the debate argues that South Africa is a country where liberty is promoted in the education system, where everyone, regardless of gender, ethnicity, culture, or language background has the right to learn. The other side contends that South Africa is a hegemonic system that facilitates success for the rich and oppresses the poor and people of color by neglecting the quality of learning experiences as a marginalized (in)social justice system. One wonders which side of this debate is true in this particular context, yet South Africa's struggle to compete against other countries for success in mathematics achievement suggests a fair and democratic education system is far from existence. How can it be that after nearly 20 years of democratic freedom, there is still isolation between functional and dysfunctional schools? How does this separation reflect a democratic ideology if the vision towards shared education success is not embraced by shared ways of teaching and learning? These questions seem to harbor an emerging sense of what we believe to be the nonappearance of metacognition, or the absence of thinking about how learners think within the curriculum.

A Legacy From Apartheid

Like much else in South Africa, the education system inherited a legacy from apartheid, which deprived many unfortunate learners of the social justice needed within the teaching and learning of mathematics. The *Curriculum and Assessment Policy Statements* (CAPS), which will be discussed in more detail in a later section of this chapter, were fully introduced in 2013 to allow for the contextual barriers and political hindrances of the past. The CAPS, therefore, served as a turning point for South Africa's education crisis, beyond the outcomes–based education dispensation. This national curriculum was realized for implementation in Grades R to 12, but not without some difficulties. Grade R is a compulsory reception year necessary to prepare all learners for schooling. The reason is that not all learners attend any pre-primary school or kindergarten. As this national curriculum was the product of 17 years of transformation in education (South Africa Department of Basic Education [SA DBE], 2012), it has been seen as a milestone after the 1994 democratic election. South Africa's curriculum was founded on the framework and guiding principles of the national Constitution (Act 108 of 1996; South Africa, Department of Justice [1996]). Through curriculum transformation, the country's education aims were modified to adhere to the following principles of the Constitution: social justice, equity, and a safe environment in which to work and learn towards building a united and democratic South Africa, able to take its rightful place across the globe as a mathematically competent country.

During the Apartheid era, some ideologies based on these principles reigned in education, including liberalism and liberation socialism (Le Grange, 2010). A geopolitical change in 1990 occurred worldwide (Le Grange, 2010) as the Post-Apartheid era paved a new beginning for South Africa and its educational system. Traditionally, the centralized curriculum model was used to develop South Africa's national curriculum statement. However, the department of education's intent for what the curriculum was supposed to be and do did not necessarily take place in practice and likely did not produce the kind of citizens desired, as evidenced by many of the applicants to universities for degrees in science, technology, mathematics, and engineering fields. Although teachers could have formed a different understanding of the curriculum's intent through their interpretation of the curriculum statements, curriculum implementation is, foremost, influenced by not only political but also philosophical (i.e., educational theories), socioeconomical, and cultural (e.g., language and milieu) influences.

Preparation for the Future: 21st Century Mathematics Education

The question we now ask is: *What preparation is required to teach for the future?* Understanding the various mathematical representations, the properties of mathematical concepts, the application of operations within certain problems, and the application of algorithms or methods—all necessitate some purpose for education towards the future. Yet, even the same content that has been taught and learned for years now seems to require different ways of teaching and learning. If we refer to the revised levels of Bloom's original taxonomy of cognitive dimensions (factual knowledge, conceptual knowledge, and procedural knowledge), we now add the metacognitive category of knowledge. Being able to use mathematical knowledge in appropriate circumstances is an essential component of mathematical proficiency in the 21st century. According to Schoenfeld (2013), there is a need for a fundamental shift from an exclusive emphasis on knowledge (what does the student know?) to a focus on what students know (and can do) with their knowledge (how? why? when?). Mathematical proficiency can, therefore, be divided into four parts: knowledge base; strategies (available and appropriate for a specific task); metacognition (using what you know effectively); and beliefs, which impact teaching and learning dispositions.

In attempting to innovate teaching and learning to prepare a new generation for the demands of this new era, many educators have discovered the value of *metacognition*, which is only implied indirectly in the CAPS (the current curriculum; SA DBE, 2012). However, Schoenfeld (2013) asserts that creating a *mathematics culture* in a classroom is the best way to develop one's metacognition. To broaden the wider mathematics education community's understandings of how this occurs in South Africa, we provide a narrative of the development of the South African mathematics curriculum, including a variety of ways that curricula are developed, understood, and implemented in different jurisdictions. First, a brief description of South Africa's education landscape is given, together with a discussion on the link between language and mathematics. This is followed by a discussion on the design principles and approaches that foster mathematical and metacognitive thinking.

CURRICULUM CHANGE: TRADITIONAL, TRANSITIONAL, AND TRANSFORMATIONAL

The following sections outline the evolution of the mathematics curriculum in South Africa. The discussion includes educational ideologies, political movements, and paradigmatic shifts that have shaped South Africa's mathematics curriculum landscape and altered assumptions of ideological absolutes.

The History of the Mathematics Curriculum in South Africa

The development of the mathematics curriculum in South Africa is synthesized in Figure 7.1. A narrative of the movements in Figure 7.1 is contained in the sections following the figure.

The Traditional Curriculum (Pre-1996)

Before 1994, South Africa hosted various decentralized education departments based on a dogma of race (such as Whites, Colored, Indian, and Black). That is, each race had its own Department of Education. The curricula were syllabus-type—the content was clearly specified with two levels of mathematics: one (higher grade) for furthering studies at tertiary level, and one for students who chose other subjects/career options (standard grade). Learners were recipients of knowledge in teacher-centered, textbook-based classrooms.

A syllabus-type interim curriculum was introduced in the years 1994 to 1996, which was in its nature an edited version of the 1983 national curriculum. All Grade 12 learners wrote, for the first time, the same examination papers (Engelbrecht & Harding, 2008). Since 1994, the education policy in South Africa has been through various development-implementation-revision cycles (SA DBE, 2014), with the assumption that systemic change requires curriculum change.

Pre-1996: **The Traditional Curriculum**
Context: Pre-1994 - different decentralized **Departments of Education** (Whites, Indian, Colored, and Black)
Content: Syllabus-type curriculum; content clearly specified with two levels of mathematics: i) one for furthering studies at tertiary level, and ii) one for learners who choose four subjects for further studies
Teacher: Traditional textbook-based talk and chalk, teachers un- or underqualified, meaning some teachers teach with grade 8 as highest qualification
Learner: Recipient of knowledge and tests and exams; performance oriented
Language of learning and teaching: Afrikaans and/or English
1994-1996 Interim curriculum: Centralized Grade 12 examination, based on previous curriculum

1997-2004: **The Transitional Curriculum (1): Outcomes-based education curriculum**
Context: Post-1994; Curriculum 2005 (Curr2005) was an outcomes-based curriculum launched and implemented since 1996/7; centralized education approach; the ideology, content, and pedagogical approach contrasted strongly with the design principles in previous and later curricula. The curriculum was open-ended; un- or under-resourced schools, such as schools in deep, rural areas with no electricity. Schooling from ages 7 to 15 or through grade 9 was compulsory.
Content: Design principle: learner- and activity-centered and resource intensive; no clear guidelines of activities, facilitating activities, or content; standards described what a learner should know and can do; not very directive and complex to implement; content not clear. The role that language plays in expression, development, and contestation of mathematics is emphasized (SA DoE, 1997); advancement of multilingualism.
Teacher: Facilitator of learning; design and develop activities for the specific needs of learners in his/her classroom
Challenges: Un- or underqualified teachers; not trained adequately for outcomes-based education
Solution: Teachers get opportunity to further qualifications and resources to be provided by department of education.
Learner: Work effectively as individuals and with others as a member of a team; solve and pose problems; language is important
Learners' under performance in international (TIMSS) and national assessments is a great concern.
2002 Due to wide criticism and challenges, this curriculum was revised.

2004-2011: **The Transitional Curriculum (2): National Curriculum Statement (NCS)**
Context: Under-resourced schools
Content: Curriculum reconstructed – simplified the outcomes statement and implemented in 2004
Design principles: Elements of teacher-centered approach; more guidelines provided; however, content still not clear; basic skills emphasized, content knowledge and grade progression added
Teacher: More direction about content and assessment; un-/under qualified teachers; teacher overload; teachers not trained adequately in outcomes-based education
Government and Department of Education provide opportunities for teachers to further/improve their qualifications.
Learner: Underperformance in international assessments but performance in national assessments (Grades 1-6 and 9) and grade 12 seems to have increased.
Considerable wide criticism of various aspects of implementation

2012-present **The Transformational Curriculum: CAPS (current curriculum in South Africa, also called the revised national curriculum statement (RNCS))**
Context: About 20 years of democracy; learners still underperforming in mathematics assessments
Content: Syllabus-type curriculum; increasingly filled with subject content and clearly specified, order of teaching the topics in each grade; provides brief summary of the principles underlying the approach, examples to clarify and descriptions of cognitive demand are provided in detail; performance oriented, but not how and when to integrate skills in subject content
Teacher: The type of teacher envisioned not mentioned; teachers have minimum opportunity for creativity, are provided with lesson plans; traditional textbook-based talk, detailed directions for tests and exams
Learner: The learner's role has shifted from a participant in the learning process and a negotiator of meaning to a recipient of pre-determined knowledge; tests and exams; performance oriented; turned back to technical, traditional teaching and learning.
*The Department of Education (DoE) was re-structured to the Department of Basic Education (DBE) and the Department of Higher Education and Training (DHET)

Fgure 7.1 Movements in South Africa's Education System.

The Transitional Curriculum (Outcomes-Based Education [Curriculum 2005], 1997–2004)

In 1997, after the first democratic election, the Department of Education (DoE) launched *Curriculum 2005*, which contrasted strongly with the curriculum in effect at the time in terms of its ideology, content, and pedagogical approach (Grussendorff, Booyse, & Burroughs, 2014). Successful performance in this outcomes-based education (OBE) approach, described as participatory, learner- and activity-centered, required the integration and application of content, competence, and confidence (Grussendorff et al., 2014). Whereas competency-based learning aims to prepare learners for success in fulfilling various life roles, mastery learning attempts to maximize the quality of teaching and the time allowed for both teaching and learning (Nair, 2003).

Teachers and learners had to take on new roles in this OBE approach. Teachers, as facilitators of learning, had to design and develop activities for the needs of their learners. Learners were required to be self-directed and self-regulated in their learning and understanding. This approach was chosen "to emancipate learners and teachers from a content-based mode of operation" (Botha, 2002, p. 5). Furthermore, this approach focused on a holistic context-appropriate set of outcomes, to develop learners' full potential in its success-orientation approach. However, teachers were not adequately trained in OBE, as they had different levels of competence to teach mathematics and design their own teaching and learning programs.

Successful performance can be achieved in an OBE approach if there is alignment between the intended outcomes, the activities that help students achieve the outcomes, and assessment of the demonstrated outcomes. Schools and teachers were given ownership of both the mathematics content of the curriculum and the process of implementation (Botha, 2002). However, many schools were under-resourced. The language of innovation associated with OBE was complex and inaccessible for many teachers, and there was concern about mathematics content in situations where un- or underprepared teachers taught mathematics by designing their own activities (Jansen, 1998). Unclear or ill-defined mathematics content requirements were present in the curriculum, and mathematics teachers were unfamiliar with the process of continuous assessment. Jansen (1998) posed many questions about OBE in early 1997, such as "How do outcomes play out in poor-resourced contexts?" (cited in Rice, 2010).[1]

Jansen (1998) believed that the broad decontextualized aims and skills of the curriculum were meaningless and would make no difference in society. Referring to the "weak culture of teaching and learning in South African schools due to under- or un-trained teachers and under-resourced schools (Botha, 2002), he was concerned about teachers' knowledge and skills to implement the OBE assessment system (cited in Rice, 2010). Many challenges were faced, such as South African learners who underperformed in international assessments such as TIMSS (South Africa's learners participated in 1995, 1999, 2002, and 2011) and national standardized assessments for Grades 1–6 and 9 (SA DBE 2011c, 2012; SA DoE, 2008; Spaull, 2013). Challenges regarding implementation of the new Language in Education

Policy were also widespread, especially in township and rural schools. This policy became a highly-contested issue within South Africa (SA DoE, 2000). Consequently, implementation of the OBE-approach to curriculum that was introduced in 1997 failed.

Reconstruction of the OBE Approach (National Curriculum Statement [NCS], 2004–2011)

In 2002, the OBE approach to curriculum was reconstructed into a National Curriculum Statement (NCS) that was implemented in 2004 (SA DBE, 2010). This revised OBE approach to national curriculum changed some design principles toward a more teacher-centered approach. The content of this new curriculum simplified the outcomes statements and provided more guidelines regarding mathematics content and assessment. Basic skills, like in OBE, were still emphasized.

Although this curriculum (NCS), introduced in 2004, generally received more positive support, there was nonetheless considerable criticism of various aspects of its implementation. These challenges included different curriculum documents that were inconsistent and therefore (a) caused confusion and stress, (b) demanded extra time from teachers that led to teacher-overload, and (c) resulted in widespread learner underperformance in international and local assessments (Grussendorff et al., 2014). Botha (2002) also noted that the school system was still characterized by a collapsed culture of teaching and learning. Being confused, overloaded, and stressed, many teachers did not know what to do when, where, and how. This meant that teachers in many schools were not trained sufficiently to teach or understand the OBE-approach (the NCS retained various dimensions of OBE); therefore, they were not equipped with sufficient knowledge and understanding of OBE, or schools were under-resourced to focus learning experiences in such a way that all learners could be successful (Botha, 2002). Thus, little teaching or learning happened, or what was taught and learned was ineffective and insufficient.

The government and DoE realized that opportunities should be provided for un- and/or under-prepared teachers to further their qualifications. Government, the DoE, and tertiary institutions worked collaboratively to provide programs in various fields for different subject needs of teachers. Implementation of the reconstructed curriculum in South African schools had a negative impact on learners who were in Grade 12 in 2005 to 2007. They were exposed to the OBE approach for five or more years and moved back to the interim (1994–1996) traditional curriculum in the last three years (2005–2007) of their schooling. The mathematics Grade 12 examination results were questioned due to the decrease in the standard of the Grade 12 Mathematics examination papers, namely that in 2008 learners passed mathematics with a 50% pass rate. It was questioned whether the standard of the 2008 results were comparable with higher grade passes in years before 2008 (Gower, 2009).

Tertiary institutions also experienced challenges. More learners passing the Grade 12 mathematics examination led to a greater number of students enrolling for mathematics courses at tertiary institutions. These students came from diverse (language and academic preparedness) backgrounds that created challenges as to the appropriate tertiary level mathematics courses at which students should start (Engelbrecht & Harding, 2008). As a response to the wide criticism on shortcomings in the implementation of the NCS over several years from a range of stakeholders, such as teachers, parents, teacher unions, school management, and academics, the second curriculum was revised again with the intention to improve the quality of teaching and learning in schools (Mouton, Louw, & Strydom, 2012).

The Transformational Curriculum (CAPS, 2012–Present, Also Called the Revised National Curriculum Statement [RNCS])

In July 2009, a panel of experts appointed by the Minister of Education investigated the nature of the challenges and problems experienced with the NCS curriculum, particularly with reference to teachers and learning quality, in order to develop a set of practical and necessary interventions in implementing the 2012 Revised National Curriculum Statement (RNCS) or CAPS (Grussendorff et al., 2014). It is important to note that the general aims, principles, and focus of the NCS remained the same in the RNCS. Once again, radical changes were planned for the implementation period from 2012–2014 (Maluleka, 2011). Changes included a reduction in the number of subjects in Grades 4–6, extended hours to focus on languages and mathematics, and a single teacher file for planning (Maluleka, 2011). The OBE curriculum included a Subject Statements file, a Learning Program Guidelines file, and a Subject Assessment Guidelines file in Grades R (Reception) to 12; teachers had to work from all these files simultaneously in the OBE curriculum. Once again, publishers had a tight schedule to revise textbooks to be ready for the 2012 implementation date (Bertram, 2011).

The amended or Revised NCS is called the CAPS (SA DBE, 2011a, b). This is the current curriculum in South Africa. The underlying framework of the CAPS curriculum for Grades R–12 is based on the following principles: social transformation; active and critical learning; high knowledge and high skills; progression (content and context of each grade shows a progression from simple to complex); and inclusivity (infusing the principles and practices of social and environmental justice and human rights as defined in the Constitution of the Republic of South Africa). The RNCS Grades R–12 (i.e., the CAPS) is still sensitive to issues of diversity, such as poverty, inequality, race, gender, language, age, disability, and other factors; valuing indigenous knowledge systems; and credibility, quality, and efficiency (SA DBE, 2011a, b). These principles, aims, and foci are only mentioned in the introductory section of the curriculum for all subjects over the education phases.

The CAPS provides detailed teaching guidelines for each content area and topic per grade per term (and per week), concepts and skills per term, clarification notes with teaching guidelines, and the duration of time allocated per topic in hours (SA DBE, 2011a, b). Teachers may choose to sequence and pace the content differently from the recommendations in this section. However, cognizance should be taken of the relative weighting and number of teaching hours of the content areas for a particular subject, phase, or grade (SA DBE, 2011a, b). The CAPS is a syllabus type curriculum that specifies the content clearly, provides workbooks for learners, and provides examples of mathematics lessons and mathematics content worked for teachers. National assessment in Grades 3, 4, 6, and 9 was introduced in this centralized education system (SA DBE, 2011c).

The CAPS stipulates the aim, scope, content, and assessment for each subject (SA DBE, 2012). The NCS Grades R–12 (SA DBE, 2012; the general aims and principles upon which the curriculum [RNCS/CAPS] was built) remained the same as in NCS, namely to produce learners who are problem solvers, who can "organise and manage themselves and their activities responsibly, and effectively; collect, analyse, organise and critically evaluate information," and communicate effectively (SA DBE, 2011a, b, p. 4). However, no guidelines are provided regarding how, where, and when instruction and integration towards aims and skills are to be organized and implemented. Therefore, one can say the aims of education have taken a back seat (Grussendorff et al., 2014).

Both mathematics teachers' and learners' roles have changed once again. According to UMALUSI (the Council for Quality Assurance in General and Further Education and Training; Grussendorff et al., 2014), the mathematics teacher's role that was clearly described in the previous curriculum is not mentioned in the CAPS introductory material. There now is silence on the role of the teacher. The learner's role in mathematics classrooms has shifted from participant in learning processes and negotiator of meaning back to recipient of pre-determined mathematics knowledge in a performance-oriented education system. In other words, the shift has been towards a more technical and traditional approach toward teaching and learning.

Support systems for mathematics Grade 12 examinations suddenly appeared. At the same time, the foundation that made the building of mathematics stronger in earlier grades and phases has been ignored or forgotten (Mouton et al., 2012). Poor mathematics Grade 12 results are a reflection of the failures of basic education and learners at the tertiary level cannot read proficiently because they failed to learn to read at the primary school level (Malada, 2010). Gower (2009) and Nel and Kistner (2009) note that university lecturers are concerned that first-year students lack basic mathematical concepts required at the school level because they (lecturers) have to re-explain mathematical procedures, such as school level equations, more than once.

Comparisons of the Outcomes-Based Education and the Current Curriculum

At this point it is insightful to compare the ideological questions regarding the OBE approach and the current mathematics curriculum, CAPS (see Table 7.1). From Table 7.1, it seems as if the design principles in the ideology in curriculum development have come full circle, currently being back at the type of education before 1994.

EDUCATION AND MATHEMATICS IN SOUTH AFRICAN SCHOOLS

To understand where these ideological positions originate, we provide some background regarding education and mathematics in South African schools. These are discussed next.

The Structure of Grades and Ages of Learners in South African Schools

In South Africa, there are primary and secondary schools. Primary schools host Grades R (Reception year) to 7, and secondary schools host Grades 8 to 12. There are four phases in schools: Foundation, Intermediate, Senior, and Further Education and Training (see Table 7.2). As shown in Table 7.2, there are five content areas in mathematics in the first nine years of

TABLE 7.1 Comparison of an Outcomes-Based Education Approach to Education and the Current Curriculum

Ideological Questions	Curr2005 (1997–2004) Towards a Transformational OBE	CAPS (2012–Present) Towards a Traditional OBE
Content (What do we teach?)	Skills, values, and knowledge	Pre-determined subject content knowledge
Method (How do we teach?)	OBE method of teaching, interactive goal-oriented method, learner- and activity-centered	Not mentioned, therefore prone to traditional lecture method
Assessment (How do we test?)	Continuous reflective process	Content assessment—tests and examinations

TABLE 7.2 Schooling Phases, Grades, Learners' Approximate Ages, Type of School, and Mathematics Content Areas

Education Band	Phase	Grades and Ages	Type of School	Mathematics Curriculum: Content Areas
General Education and Training (GET)	Foundation	Grades R–3 Ages 6–10	Primary	1. Numbers, Operations, and Relationships 2. Patterns, Functions, and Algebra 3. Space and Shape (Geometry) 4. Measurement 5. Data handling
	Intermediate	Grades 4–6 Ages 9–14		
Compulsory education for all learners	Senior	Grades 7–9 Ages 13–17	Primary: Grade 7	
			Secondary: Grades 8–9	
Further Education and Training (FET)	Further Education and Training (Excluding FET-Colleges)	Grades 10–12 Ages 15–19	Secondary	1. Functions 2. Number patterns, sequences, and series 3. Finance, growth, and decay 4. Algebra 5. Differential calculus 6. Probability 7. Euclidean geometry and measurement 8. Analytical geometry 9. Trigonometry 10. Statistics

Source: SA DBE (2011a, b).

schooling (Foundation to Senior phase). These content areas form a learning trajectory, building on previous knowledge of topics/themes over grades and phases. The five content areas evolve into ten mathematics content areas for the last three years of schooling (Grades 10 to 12).

The Development of the Definition of Mathematics in South Africa

Amidst contextual challenges, it has become evident that the mathematics school curriculum in South Africa is indeed a product of socialism. We explain this in Table 7.3, which provides two definitions of mathematics as a cultural tool in the traditional OBE curriculum (1994 curriculum & SA DoE, 1997) and more recent CAPS curriculum (SA DBE, 2011a, b). The differences and similarities between these two views highlight language as a tool for communication, thinking, and politics in, with, and about mathematics.

Language Complexity in the South African Mathematics Curriculum

Parallel to the development of education policy in South Africa, one also has to consider the development and implementation of language in education policy (LiEP). The advancement of multilingualism (SA DoE,

TABLE 7.3 Definitions of Mathematics in Two South African Curriculum Documents

OBE Definition of Mathematics (SA DoE, 1997)	CAPS Definition of Mathematics (SA DBE 2011a, b)
Mathematics is the construction of knowledge that deals with qualitative and quantitative relationships of space and time. It is a *human activity* that deals with patterns, problem solving, logical thinking, etc., in an attempt to understand the world and make use of that understanding. This understanding is expressed, developed, and contested through *language, symbols, and social interaction*.	*Mathematics is a language* that makes use of symbols and notations for describing numerical, geometric, and graphical relationships. It is a *human activity* that involves observing, representing, and investigating patterns and qualitative relationships in physical and social phenomena and between mathematical objects themselves. It helps to develop mental processes that enhance logical and critical thinking, accuracy and problem solving that will contribute in decision-making.

Source: Based on SA DoE (1997) and SA DBE (2011a, b).

1996) is part of the democratic Constitution of South Africa. Before 1994, a controversial system of language policy in African schools had developed over many years. Learners received mother tongue education in the early years of schooling, but had to change to English and/or Afrikaans later in their schooling.

Education in South Africa was initiated in the late 17th to the 19th centuries by missionaries, who used English as the medium of instruction. As a British colony (1910 to 1948), the Union of South Africa government decentralized decisions about the language of learning and teaching (LOLT) in schools to the different provinces and departments of education (Beukes, 2004; Hartshorne, 1987).

Since the early 20th century, English and Afrikaans have been the only official languages. A decentralized, segregated education system based on race, provinces, and homelands left the responsibility to make decisions regarding issues, such as the language of learning and teaching, to provinces and departments of education and homelands.

After 1994, the democratic South African government centralized the education system. The Constitution of the Republic of South Africa (South Africa, 1996) identified 11 official languages of the country: Sepedi, Sesotho, Setswana, Siswati, Tshivenda, Xitsonga, Afrikaans, English, isiNdebele, isiXhosa, and isiZulu. Since that time, learners (or parents) have been afforded the right to study in the language of their choice, with the intention that the medium of instruction be the mother tongue in the first four years of schooling. This choice is indicated by learners/parents during the application for admission to a particular school. Despite the provision in the Constitution of South Africa that all official languages must enjoy parity of esteem, the home language gradually shifted from the mother tongue as people started to move towards uni-lingualism, especially in urban and metropolitan sites (Mouton et al., 2012). English, as the language of power, politics, government, economics, and media, was increasingly adopted as the home language (though it was not the mother tongue). As a consequence, many schools adopted English as the language of teaching and learning (Taylor & Vinjevold, 1999). Thus, schools in South Africa have more recently become identified as parallel, dual, or single medium schools for teaching and learning. *Parallel medium* means teaching using two languages of instruction in separate classrooms. *Dual medium* means using two languages of instruction, switching (repeating) from one to the other in one classroom. *Single medium* schools use one language (either English or Afrikaans) as the language of learning and teaching. *Code-switching*, an alternative medium of teaching and learning, means switching from one language of instruction to an African language (that is not a language of instruction, but is many learners' mother tongue).

According to Webb (2002), language plays a central role, as a medium of learning, in educational development. Setati (2002a) adds that "mathematical speech and writing have a variety of language types that learners need to understand in order to participate appropriately in any mathematical conversation" (p. 10). According to Setati (2002b), the teacher is the mediator between mathematics and learners when learners engage in sharing ideas and reflecting upon their understanding of mathematics concepts. However, the acceptance of all 11 official languages as languages of learning and teaching has far reaching challenges, both enabling and constraining the teaching and learning of mathematics in schools. Setati and Adler (2001) present a challenge: Imagine that "both teachers and learners are multilingual, but none have English, as their main language." What happens in such classrooms is that teachers teach using code-switching, but again, not all learners in the classroom necessarily have one of these languages as their main language. Tests and examinations are written in either English or Afrikaans, which can be learners' third or fourth language. This situation is far from acceptable. It is also important to note that none of the African languages has been developed and established as an academic language, even after 20 years under a democratic government.

Media Reports About Mathematics Teaching and Learning in South Africa

There have been numerous reports on education in *The Mail & Guardian*, an accredited newspaper in South Africa. Henning (*Mail & Guardian*, 18 Jan 2013) reported that economists are concerned about mathematics because "they think of the mathematics skills needed to run a prosperous economy" and that parents "hunt for schools where their children may get a better deal." The economists (Graven, 2015) also referred to how the "systemic red tape impacted on how teachers taught and how much time they had available to undergo professional development courses."

After the Grade 12 results were released, education reporter Nkosi (2015) noted some serious challenges that the new curriculum (CAPS) had presented, including the fact that the pass percentage of mathematics in Grade 12 had declined significantly, from 59.1% to 53.5%. Furthermore, only 3.2%, or 7,216 of Grade 12 students achieved 75% or higher in mathematics. According to Nkosi (2015), there are positive perceptions about the future of the CAPS, provided the curriculum is not revised/reconstructed again, as revising might create a repeat of the confusion teachers had with the curriculum's terminology (e.g., objective versus outcome).

Teacher unions, such as the South African Democratic Teacher Union, complained that teachers had not been thoroughly trained to teach the new curriculum, and recommended that training or professional development should precede and accompany piloting and implementing changes in the curriculum. Another teacher union, the National Association of Professional Teachers of South Africa, focused attention on the distressing fact that "those who wrote the matric exam in 2014, are just 42.5% of the cohort of 1,252,071 pupils who enrolled in Grade 1 in 2003," which was the OBE-era (Nkosi, 2015). Thus, student retention is of great concern in South African schools.

Significant progress has been made relative to administrative restructuring, policy development, and infrastructural improvement, but the quality of (mathematics) education is for most learners far from satisfactory (Howie, 2004). One can conclude that curriculum development still needs major changes to address the many serious challenges faced in education. However, we argue that changes should not be made in the same way that has been used during the first 20 years of democracy in South Africa. Instead, we suggest the fostering of a different kind of curriculum, one that emphasizes metacognitive thinking.

METCOGNITION AS AN EMERGING IDEOLOGY IN CURRICULUM

A proposed metacognitive ideology is, we argue, the key set of beliefs and values that underscore the reconstruction needed in the curriculum, as typical of a curriculum in a constant revision cycle. We contend that a curriculum needs to serve the interests of learners as complete human beings, developing their full potential (mentally, spiritually, and physically) to emancipate their economic potential and their sensitivity to different cultural and language backgrounds. We argue for a more permissive movement, away from isolated disciplines and hierarchical relations between teachers and learners, to a more open, liberal learning experience that will empower awareness of the value added if metacognition is included as one of the foci in mathematics education. This should be regarded as a holistic and value driven view of the teaching and learning of mathematics, a view that fosters the awareness of self and others' strengths and weaknesses in the teaching and learning of mathematics.

We suggest that the movement towards a transcending metacognitive ideology should be based on a dimension of *metacognitive knowledge*, which refers to the strategic or reflective knowledge about how to go about solving problems and cognitive tasks, to include contextual and conditional knowledge, and knowledge of self. Although metacognition is referred to indirectly in the CAPS, it is important that mathematics is learned on all cognitive levels (Du Plooy & Long, 2014). In the absence of metacognition, the understanding of mathematics turns to memorizing facts and procedures, without a deep sense of awareness of the value of learning mathematics.

Human, van der Walt, and Posthuma (2015) associated the CAPS and mathematics ideologies as indicated in Table 7.4. It seems that the CAPS teacher can be classified as both that of a progressive and a public educator, each with specific principles and approaches that contribute to the ideologies.

DISCUSSION OF THE TRANSCENDING METACOGNITIVE IDEOLOGY

Higher education institutions' main aim for teaching-learning in the 21st century is to improve metacognitive competencies that enhance students' ability to formulate their learning goals, evaluate their learning outcomes, and manage their own cognition and actions appropriately to become intentional lifelong learners (Savin-Baden & Major, 2004). Metacognitive processes are viewed as a prerequisite of lifelong self-directed learning (Garrison, 2003).

Learning how to learn is mentioned directly and indirectly in the introduction pages to the CAPS documents (SA DBE, 2011a, b). The principles the NCS (also RNCS) are based on include "active and critical learning: encouraging an active and critical approach to learning, rather than rote and uncritical learning of given truths; high knowledge and high skills: the minimum standards of knowledge and skills to be achieved at each grade are specified and set high, achievable standards in all subjects" (SA DBE, 2011a, b, p. 5). However, these principles are not referred to or explained where the content is provided and explained, and therefore, it is implemented ineffectively at both school and tertiary levels. Perkins and Wirth (2008) propose the addition of a thinking- or meta-curriculum to deepen understanding of subject content knowledge and to develop the skills necessary to apply these in students' understanding of complex real-world problems. For instance, students' keeping learning journals facilitates reflective practice in and on action and builds metacognitive capacities (Institute for Adult Learning, 2012). Reflective approaches turn students into active learners, provide for learning from the process, and support the planning, monitoring, and evaluating of progress in a project (problem solving; Moos, 2014).

Our intention with this chapter was to describe the development of the mathematics curriculum in South Africa and to identify the need for a metacognitive ideology as a transcending construct in the mathematics

TABLE 7.4 A Comparison of Mathematics Education Ideologies and CAPS in Terms of Social Elements

Social Elements of the Ideology	Progressive Educator Ideology	Public Educator Ideology	CAPS (SA Curriculum)
View of mathematics	Process view Personalized mathematics Language and human activity	Social constructivism	Unique language Human activity Social constructing of mathematical ideas and concepts
View of the child	Child-centered Progressive view Child viewed as a growing flower and innocent savage	Social conditions view the child as "clay moulded by environment" and "sleeping giant"	Learner-centered Promote holistic development Progression from one grade to the next Social conditions
View of ability	Abilities vary but need cherishing		Differentiated activities according to each learner's ability
Mathematical aims	Creativity, self-realisation through mathematics (child-centered)	Critical awareness and democratic citizenship via mathematics	Self-realisation through mathematics Confidence and competence to handle any mathematics situation Creative activity Critical awareness of the role of mathematics in society, environments, cultures and economies
Theory of learning	Activity, play, exploration	Questioning, decision making, negotiation	Play, develop understanding of number and numeracy Interactive Do, speak, demonstrate Develop mathematical thinking
Theory of teaching mathematics	Facilitating personal exploration, preventing failure	Discussion, conflict, questioning of content and pedagogy	Integrated approach Learn through play Facilitator of learning
Theory of resources	Rich environment to explore		Group work Discussions
Theory of assessment in mathematics	Teacher-led internal assessment, avoiding failure	Various modes Use of social issues and content	Various methods Teacher-led internal assessment Grade 9 external assessment

Source: Human, van der Walt, & Posthuma (2015, p. 4)

curriculum. The nature of the relationship between language, curriculum content, and metacognition was elaborated in the theoretical framework to yield a theoretical view of the metacognitive ideology. As explained, reflection on experiences with mathematics creates awareness of the affective, metacognitive, and meta-affective experiences that shape one's worldview, and thus the ideology of metacognition. Figure 7.2 illustrates the three levels that constitute this metacognitive ideology. In this way, beliefs, intentions, attitudes, and experiences supplement the mathematical content that resides in the curriculum and, at the same time, demand particular teaching skills, especially planning, monitoring, and evaluation. Furthermore, knowledge of one's own and others' cognition, cognitive demands of tasks, and available and appropriate strategies, as well as procedural, conceptual, and declarative knowledge, are part of teaching and learning of and in mathematics.

Typical of mathematics learning, problem-solving experiences are implicit in nature and, when reflected upon, create awareness of the affective, metacognitive, and meta-affective experiences on a perceptual level (Jagals, 2015). We view the metacognitive ideology as a perceptual ideology, one that emerges from our own intentions, beliefs, attitudes, and experiences with curriculum change in South Africa. To put it differently, a transcending metacognitive ideology emerges as an interpretation of curriculum

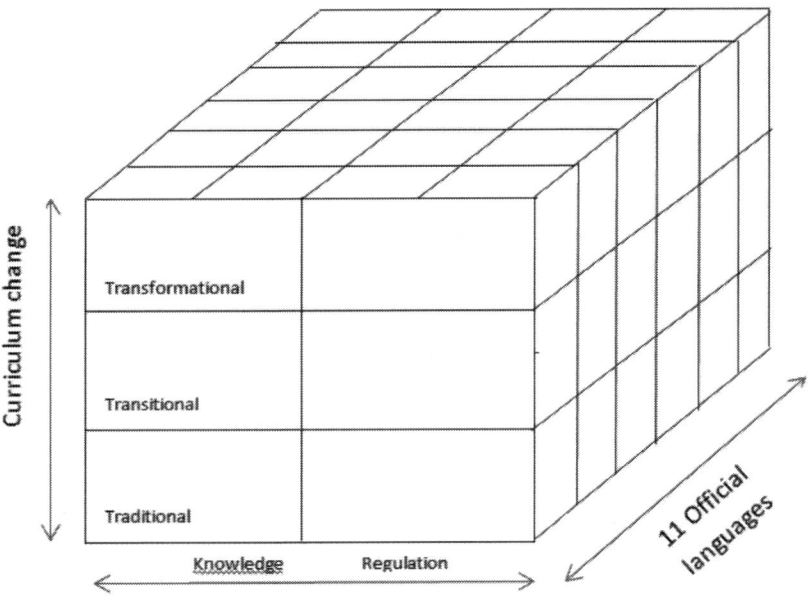

Figure 7.2 Taxonomy of the metacognitive ideology in the mathematics curriculum.

change and is based on change for, of, and with metacognitive thinking as an emerging construct. The taxonomy in Figure 7.2 depicts how the official languages of South Africa necessitate awareness in curriculum change to adopt a curriculum developmental view that reflects the needs of all learners, to embrace their understanding through their language. We say this because language interacts with metacognitive thinking, and impacts knowledge of ourselves and others. Furthermore, the transcendence to a more transformational curriculum will require not just content knowledge, but metacognitive awareness of the knowledge and regulatory skills needed to acquire and develop mathematical understanding. This taxonomy shows the developmental level of metacognition as an implicit experience level of knowing assumed by a transformational curriculum, such as anticipated by the CAPS curriculum in South Africa.

CONCLUSION AND FUTURE DIRECTIONS

Curriculum change allows us to distinguish between the robust and the frail of, what we believe to be, normality and superlative. Curriculum that has withstood the threshold of time is, normally, what we regard as best plans of practice for a specific context. These best plans are often used to locate possible reasons for what, we believe, leads to success in education. To develop new plans for the future classroom, curriculum developers and teachers need to consider thinking about how future societies will (learn to) reflect—that is, how they will think metacognitively, and instill this within the curriculum. Towards this end, the curriculum often transcribes a plea for a dogma of pure, balanced, and ethical righteousness and seems driven by the principles of democracy. This all exists in some present space within the foundations of the curriculum, and serves to protect its citizens and offer them hope for tomorrow. If the curriculum's intent gets lost within a false pedagogy, within a strive for survival and peer competence, within politics and disputes, within economic fallacies, we could very well close the road to tomorrow, as the foundations upon which it rests may not be as sturdy as intended by curriculum theorists' beliefs. We would then be left searching between the rubble of lost spaces of past curricula for a guide or map of some sort; to start anew, we need to rethink where we ought to be and consider the consequences. We would then think of the cultural, spiritual, emotional, and academic spaces of future societies, and what we need to consider building those spaces, to occupy them with intended citizens. We will, then, need a blueprint of what these spaces might look like to plan and monitor the road ahead. To do so, we revisit the avenues of past curricula and wander about them with a sense of objectivity to revise their intent for society and its future. These spaces of past, present, and future reflections could provide the freedom we need to deliver a curriculum for the future. Yet, to do that, we need to know where the present comes from, where it is located, and where the road is taking us. Only then can we draft a blueprint of tomorrow's curriculum to cater for the future spaces we have yet to live in.

One concern that is illuminated by this chapter is the relatively weak link between the philosophy of education and mathematics education. This is a relationship that needs to be more actively explored and developed. Like mathematics, mathematics education has too much of a tendency to isolate it from adjacent areas of knowledge and inquiry. How often, for example, do we look to developments in science education? The philosophy of mathematics education is one approach that should facilitate building links with other areas of knowledge and research. This chapter suggests a number of directions in which we could turn our gaze to begin to do this.

What still needs to be discussed in mathematics education research and included in curriculum policy are multi-lingualism and the language of the country. South Africa has 11 official languages. The language of learning and teaching is decided by the school governing body and, in most cases, learners are taught in a second, third, or even fourth language. Further discussion is also needed on the urbanization and decolonization of the curriculum to infuse indigenous knowledge systems with westernized mathematics concepts to teach in culturally receptive ways.

NOTE

1. This is an updated, electronic version of Jansen's original paper: "Why OBE will Fail," first presented at a conference at the University of Durban Westville in 1997.

REFERENCES

Bertram, C. (2011, June 9). Rushing curriculum reform again. *Mail and Guardian*, p. 39.
Beukes, A. (2004, Month, May 20–23). *The first ten years of democracy: Language policy in South Africa*. Paper presented at the 10th Linguapax Congress on Linguistic Diversity, Sustainability and Peace, Barcelona, Spain.
Botha, R. J. (2002, March). *The introduction of a system of OBE in South Africa: Transforming and empowering a marginalized and disenfranchised society*. Paper presented at the Annual Meeting of the Comparative and International Education Society, Orlando, FL.
Du Plooy, R., & Long, C. (2014, July). Engaging with cognitive levels: a practical approach towards assessing the cognitive spectrum in mathematics. *AMESA*

Conference Proceedings. Retrieved from http://www.amesa.org.za/AMESA2014/Proceedings/papers/Short%20Paper/2.%20Ryna%20du%20Plooy.pdf

Engelbrecht, J., & Harding A. (2008). The impact of the transition to outcomes-based teaching on university preparedness in mathematics in South Africa. *Mathematics Education Research Journal, 20*(2), 57–70.

Ernest, P. (1991). *The philosophy of mathematics education*. Hampshire, England: The Falmer Press.

Ernest, P. (2010). Reflections on theories of learning. In B. Sriraman & L. English (Eds.), *Theories of mathematics education: Seeking new frontiers* (pp. 39–47). Zürich, Germany: Springer.

Flavell, J. H. (1979). Metacognition and cognitive monitoring: A new area of cognitive-developmental inquiry. *American Psychologist, 34*(10), 906–911.

Garrison, D. R. (2003). Self-directed learning and distance education. In M. G. Moore & W. Anderson (Eds.), *Handbook of distance education* (pp. 161–168). Mahwah, NJ: Lawrence Erlbaum.

Gower, P. (2009, May 1). Maths exam finding. *The Teacher*, p. 5.

Graven, M. (2015, 27 March). What's wrong with maths? *Mail & Guardian*. Retrieved from https://mg.co.za/article/2015-03-27-00-whats-wrong-with-maths-today

Grussendorf, S., Booyse, C., & Burroughs, E. (2014). *What's in the CAPS package? Report Overview*. Retrieved from http://www.umalusi.org.za/docs/reports/2014/overview_comparitive_analysis.pdf

Hartshorne, K. B. (1987). Language policy in African education in South Africa, 1910–1985, with particular reference to the issue of medium of instruction. In D. Young (Ed.), *Bridging the gap between theory and practice in English second language teaching* (pp. 62–81). Cape Town, South Africa: Maskew Miller Longman.

Henning, E. (2013, 18 January). Solving the problem of Maths. *Mail & Guardian*, Retrieved from https://mg.co.za/article/2013-01-18-solving-the-problem-of-maths.

Howie, S. (2004). A national assessment in mathematics within an international comparative assessment. *Perspectives in Education, 22*(2), 149–161.

Human, A., van der Walt, M., & Posthuma, B. (2015). International comparisons of foundation phase number domain mathematics knowledge and practice standards. *South African Journal of Education, 35*(1), 1–13.

Institute for Adult Learning. (2012). *Tools for re-imagining learning: Understanding metacognition*. Retrieved from https://www.ial.edu.sg/files/documents/457/TLD%20Report_External%20Report.pdf

Jagals, D. (2015). *Metacognitive locale: a design-based theory of students' metacognitive language and networking in mathematics*. Unpublished PhD thesis, North-West University Potchefstroom Campus, South Africa.

Jansen, J. (1998). *Curriculum reform in South Africa: A critical analysis of outcomes-based education*. Retrieved from http://repository.up.ac.za/bitstream/handle/2263/132/Jansen%20(1998)a.pdf

Le Grange, L. (2010). African curriculum studies, continental overview. In C. Kridel (Ed.), *Encyclopedia of curriculum studies* (Volume 1, pp. 18–22). New York, NY: SAGE. Retrieved from http://dx.doi.org/10.4135/9781412958806.n10

Malada, B. (2010, September 19). We ignore proper education at our peril. *Sundae Tribune*, p. 22.

Maluleka, S. (2011, August 10). Curriculum: Back to the blackboard. *Daily News*, p. 3.

Moos, D. C. (2014). Setting the stage for the metacognition during hypermedia learning: What motivation constructs matter? *Computers & Education, 70*, 128–137.

Mouton, A. N., Louw, G. P., & Strydom, G. L. (2012). Historical analysis of the Post-Apartheid dispensation education in South Africa (1994–2011). *International Business & Economics Research Journal, 11*(11), 1211–1221.

Nair, P. A. P. (2003). Can prior learning experience serve as a catalyst in the paradigm shift from traditional teaching methodology to outcomes-based educational practice? Perspectives on higher education. *South African Journal of Higher Education, 17*(2), 68–78.

Nel, C., & Kistner, L. (2009). The National Senior Certificate: Implications for access to higher education. *South African Journal of Higher Education, 23*(5), 953–973.

Nkosi B. (2015, 6 January). Matric pass-rate drop disappointing but understandable. *Mail & Guardian*. Retrieved from https://mg.co.za/article/2015-01-06-matric-pass-rate-drop-disappointing-but-understandable

Perkins, D. & Wirth, K. R. (2008). *Learning to learn* [PowerPoint presentation]. Available from http://www.macalester.edu/geology/wirth/CourseMaterials.html

Rice, A. (2010, July 7). *Analysis: RIP outcomes-based education and don't come back*. Retrieved from https://www.dailymaverick.co.za/article/2010-07-07-analysis-rip-outcomes-based-education-and-dont-come-back

Samuelsson, M., & Samuelsson, J. (2016). Gender differences in boys' and girls' perception of teaching and learning mathematics. *Open Review of Educational Research, 3*(1), 18–34.

Savin-Baden, M., & Major, C. H. (2004). *Foundations of problem-based learning*. London, England: McGraw-Hill Education.

Schoenfeld, A. H. (2013). Reflections on problem solving theory and practice. *The Mathematics Enthusiast, 10*(1–2), 9–34.

Setati, M. (2002a). Researching mathematics education and language in multilingual South Africa. *The Mathematics Educator, 2*(2), 6–20.

Setati, M. (2002b). *Language practices in intermediate multilingual mathematics classrooms* [Unpublished doctoral dissertation]. University of the Witwatersrand, Johannesburg, South Africa.

Setati, M., & Adler, J. (2001). Between languages and discourses: Language practices in primary multilingual mathematics classrooms in South Africa. *Educational Studies in Mathematics, 43*(3), 243–269.

South Africa, Department of Justice. (1996). *The Constitution of the Republic of South Africa*. Retrieved from http://www.justice.gov.za/legislation.constitution/SAConstitution-web-eng.pdf

South Africa Department of Basic Education [SA DBE]. (2010). *Report of the ministerial committee for LTSM*. Pretoria, South Africa: DBE.

South Africa Department of Basic Education [SA DBE]. (2011a). *Curriculum and assessment policy statement GRADES 10-12 MATHEMATICS*. Pretoria, South Africa: DBE.

South Africa Department of Basic Education [SA DBE]. (2011b). *National curriculum statement (NCS): Curriculum and assessment policy statement Intermediate Phase Grades 4-6.* Pretoria, South Africa: DBE.

South Africa Department of Basic Education [SA DBE]. (2011c). *Report on the annual national assessments of 2011.* Pretoria, South Africa: DBE.

South Africa Department of Basic Education [SA DBE]. (2012). *Report on the annual national assessments 2012.* Pretoria, South Africa: DBE.

South Africa Department of Basic Education [SA DBE]. (2014). *Report on the annual national assessments of 2014 Grades 1 to 6 and 9.* Pretoria, South Africa: DBE

South Africa Department of Education [SA DoE]. (1996). *Curriculum framework for GET and FET.* Pretoria, South Africa: National Department of Education (NDE).

South Africa Department of Education [SA DoE]. (1997). *Senior phase policy document.* Pretoria, South Africa: NDE.

South Africa Department of Education [SA DoE]. (2000). *National centre for curriculum and research language in the classrooms: Towards a framework for intervention.* Pretoria, South Africa: NDE.

South Africa Department of Education [SA DoE]. (2008). *Grade 3 systemic evaluation 2007.* Leaflet. Pretoria, South Africa: DoE.

Spaull, N. (2013). Skills and education—Accountability in South African Education. In N. Spaull (Ed.), *2013 South Africa's education crisis: The quality of education in South Africa 1994-2011.* Report Commissioned by Centre for Development and Enterprise. Pretoria, South Africa: CDE.

Taylor, N., & Vinjevold, P. (1999). *Getting learning right.* Johannesburg, South Africa: Joint Education Trust.

Webb, V. (2002). *Language in South Africa. The role of language in national transformation, reconstruction and development.* Amsterdam, the Netherlands/New York, NY: John Benjamin.

CHAPTER 8

DISCUSSING THE MATHEMATICS CURRICULUM IN BRAZIL

Celi Espasandin Lopes
Universidade Cruzeiro do Sul

Regina Célia Grando
Universidade Federal de Santa Catarina

This chapter describes the mathematics curriculum for Brazilian basic education. In Brazil, basic education is the first level of schooling, which is comprised of three stages: early-childhood education (for students aged 0 to 5), elementary school (for students aged 6 to 14), and high school (for students aged 15 to 17).

This education of children and youth is considered a universal right of individuals and an indispensable foundation for exercising full citizenship. It is the time, space, and context in which individuals shape and reshape their identity, by learning to respect and value diversity amidst physical, affective, emotional, socio-emotional, cognitive, and cultural changes (Ministério da Educação e do Desporto do Brasil [MEC], 2013). Freedom and pluralism are, therefore, prime requirements of the educational system and are central principles for the vision of mathematics education as well as for designing curricular guidelines.

In this chapter, we initially present a historical perspective of the Brazilian curriculum and guidelines for teaching mathematics. We then describe the context of Brazilian education and the influences exerted by public policy on curricular development. After that, we highlight the underlying structure and organization of the curricular guidelines for basic education, in order to characterize and discuss the vision of mathematics education portrayed in the documents, and the incorporation of indicators arising from research about the mathematics curriculum. We also consider issues under discussion in the Brazilian educational scenario as regards the perception that parents, teachers, and students have about the curriculum, in addition to the challenges faced during curriculum implementation.

HISTORICAL PERSPECTIVE OF THE BRAZILIAN MATHEMATICS CURRICULUM AND TEACHING GUIDELINES

The development of the Brazilian curriculum, at the national level, has always been the responsibility of the Ministry of Education, an agency that hires experts who write curriculum documents according to educational legislation. The Brazilian educational tradition regarding the curriculum is based on prescriptive documents, which define disciplines, content topics, workload, teaching methods and techniques, as well as examinations of pre-established goals. However, the official curriculum in these documents is often disconnected from what actually goes on in Brazilian classrooms, that is, from the implemented curriculum (Lopes, 1998). This is due to cultural, social, geographic, and economic differences that are realities in 26 Brazilian states, distributed in five regions. The northeast region has nine states, the Southeast region has four states, the southern region has three states, the Midwest has three states and one federal district, and the northern region has seven states.

The study of curriculum development in Brazil shows that curriculum decisions have historically been marked by governmental actions and have not arisen from movements originating in schools, spearheaded by teachers, or civil society. Concerning curriculum issues, one of the hallmarks of Brazilian public policy is the lack of curriculum implementation actions, as if new ideas would automatically turn into practice (Barretto, 2012). Besides this lack of implementation action, another attribute of such policies is the lack of monitoring of the proposed innovations, which compromises proper evaluation.

There is an understanding that the control of curriculum implementation actions and student performance regarding the changes must be done through external evaluations by federal, state, and local governments. This creates an overload of assessments whose outcome causes little impact on public policy. In the case of the mathematics curriculum, a prime example happened in the 1930s when Professor Euclides Roxo proposed the unification of the fields of Algebra, Arithmetic, and Geometry into the single discipline of Mathematics to address all of the strands in an interrelated manner (Pires, 2008). There was also a recommendation that a practical, hands-on approach to geometry should precede the introduction to deductive geometry. This view of the curriculum was expanded beyond mere lists of content to be taught to include a discussion of didactic guidelines. However, it did not appear that this significant achievement was consolidated in the curriculum reform that followed.

In the period from 1960 to 1980, the Modern Mathematics Movement, introducing topics related to Set Theory and emphasizing the use of symbolic language, influenced the Brazilian mathematics curriculum. The period was marked by legislation that attributed the responsibility of developing curricula to each of the states of the federation (D'Ambrosio, 1987). Official curricula gradually incorporated the ideas of the Modern Mathematics Movement. Even before the government agencies formalized the curriculum guidelines, the collections of textbooks sold in the country had introduced themes related to Set Theory, emphasizing the use of symbolic language. Because textbooks propagated such ideas, these ideas were quickly adopted by the schools, and exerted great influence, especially in the selection of teaching content. Beginning in the 1980s, the consolidation of "Mathematics Education" as an area of theoretical and practical research in Brazil brought contributions in relation to the content to be taught, why teach it, how to teach it, and how to assess learning (Miguel, Garnica, Igliori, & D'Ambrosio, 2004).

In 1996, a bill was signed into law, which stated that the federal government, in cooperation with states and cities, would set guidelines for the development of the curriculum to ensure a common basic education. This legal provision caused the Brazilian Ministry of Education to prepare a document called *Parâmetros Curriculares Nacionais* [National Curriculum Parameters] (PCN; MEC, 1997) for different levels and types of education. The task of devising national requirements implied coping with many concerns and with the need to address questions, such as how to create national references to tackle long-standing problems of Brazilian education. At the same time, new challenges were posed because of the global scenario as well as new characteristics of society, such as increasing urbanization. Moreover, Brazil was challenged with how to establish standardized national guidelines for a country with continental dimensions and enormous diversity across the country.

Brazil is considered a country with continental dimensions because of its territorial extension of 8,514,876 km^2. Its area corresponds to approximately 1.6% of the entire surface of the planet, occupying 5.6% of the land area of the globe, 20.8% of the area of all the Americas, and 48% in South America. It is the fifth largest country on the planet. The large size of Brazil gives the country a huge diversity of landscapes, climates, topography, flora, and fauna. This territorial extension and the geographical differences tend to be complicating factors for the implementation of public policies, particularly those related to curriculum implementation. Regarding the mathematics curricula, the preparation of the PCN achieved consensus from the community of mathematics educators who believed that their most significant demands had been integrated into the document.

The document clarified the role of mathematics in basic education by proposing goals, which showed students the importance of viewing mathematics as an instrument to understand the world around them, and see it as a field of knowledge that stimulates interest, curiosity, and an inquisitive spirit, as well as nurturing the ability to solve problems. In addition, it emphasized the importance of students developing confident attitudes towards their own ability to construct mathematical knowledge, cultivating self-esteem, respecting the work of colleagues, and persevering in the quest for solutions. This curricular perspective adopted social relevance and the contribution to the intellectual development of students as selection criteria for content (MEC, 1997).

The selection criteria were an innovation, as they present content not only from a conceptual dimension, but also in relation to procedures and attitudes. The criteria emphasized the importance of overcoming the linear organization of content and the need to demonstrate connections among content topics. Moreover, the criteria incorporated directions arising from research in mathematics education regarding problem-solving, the history of mathematics, Ethnomathematics, and mathematical modeling, as well as information and communication technologies.

The *Parâmetros Curriculares Nacionais para o Ensino Médio*—[National Curriculum Parameters for Secondary Education] (PCNEM; MEC, 2000) focused on contextualization and the preparation for work; it centered on contextualized mathematics as an instrument for functioning in the workforce. In contrast, the *Referencial Curricular Nacional para a Educação Infantil* [National Curriculum Framework for Early-Childhood Education] (RCNEI; MEC, 1998) focused on mathematical concepts that can be developed in childhood, especially through recreational activities.

The dialogue with researchers in mathematics education and with the community of mathematics educators helped to prepare the document

Orientações Curriculares para o Ensino Médio [Curricular Directives for Secondary Education] (OCEM; Lopes, Santos, Gravina, & Carvalho, 2006). Representatives of students, teachers, and school management throughout Brazil discussed this document, which was later organized by four distinguished university professors from different parts of the country who had extensive teacher training experience as well as experience with high school teachers. Lopes, Santos, Gravina, & Carvalho (2006) expanded the ideas of mathematics teaching and learning in light of new technologies, mathematical modeling, historical contextualization, project work, and literacy. Moreover, rather than just listing content, they discussed the mathematical concepts and procedures that merited emphasis in high school classrooms.

However, all of these documents eventually became mere pieces of paper. Due to the size of the Brazilian territory, such documents were hardly discussed in schools, and ultimately became disconnected from the reality of the classrooms. In the opinion of many Brazilian mathematics educators, the documents PCN, RCNEI, and PCNEM filled the gap, though misguidedly and with little dialogue, by defining the methodological concepts to be used as well as the mathematical content to be taught in kindergarten, elementary, middle and high school. Curricular Directives for High School, despite resulting from a comprehensive national debate and significant involvement of members of *Sociedade Brasileira de Educação Matemática* [the Brazilian Mathematics Education Society] (SBEM) and the *Sociedade Brasileira de Matemática* [the Brazilian Mathematics Society] (SBM), ended up trapped in officialdom. That is, the documents were viewed as official documents but without influence on the classroom.

Elaborated in different forms, such documents were seen as mandatory content requirements to be taught throughout Brazil, as if they were a script, suggesting that such measures would be sufficient direction to ensure quality education for all. However, universal education cannot be accomplished by decree, ordinance, resolution, or any similar official act. It is not attained merely through prescription of teaching activities, or the establishment of curricular parameters or guidelines. Education with social value is an accomplishment, and as an accomplishment of Brazilian society, it must be manifested through social movements as a universal right. Social stakeholders must take ownership of curriculum documents. They need to understand the pedagogical directives and the different areas of mathematics education contemplated, as well as translate these ideas into curriculum practices in the mathematics classroom.

This disconnect between the official and implemented curriculum has been corroborated by research evidence regarding the Brazilian educational system, which shows that schools differ not only in the diversity of pedagogical and administrative approaches adopted, but mainly in the variety of internal practices and structures, such as the environment, the level of teachers' commitment, and the emphasis given to teaching and learning processes. "School makes a difference," and in a country characterized by clear socio-cultural differences, public policies should pay closer attention to the differences among schools (Garnica, 2014, p. 148).

Thus, the development of curriculum guidelines needs to take into account such aspects, and enable schools to participate more effectively in the definition of their reality, as well as develop proposals that emerge from their peculiarities and complexity (Lopes, 1998).

Current Principles of Brazilian Education and Curriculum Organization

Brazil's recent educational scenario has required large investments in public policy. The 2010 School Census, conducted by *Instituto Nacional de Estudos e Pesquisas Educacionais Anísio Teixeira* (INEP, 2014), showed that currently Brazil has 51.5 million students enrolled in both private and public basic education institutions, at early-childhood, elementary, and high school levels, vocational and special needs institutions, as well as young adult and adult education institutions. Of these 51.5 million, 43.9 are enrolled in public institutions (85.4%) and 7.5 million in private schools (14.6%).

These figures constitute a major challenge for a country that is still struggling to ensure the access and permanence of children in basic education. In 2015, the Ministry of Education, reacting to this scenario and in accordance with the National Education Plan, developed a curriculum document called *Base Nacional Comum Curricular* [Common National Curriculum Base] (BNCC), currently under Public Inquiry from scientific societies, individual schools, academia, and the general population. The objective of BNCC is to indicate learning and development paths for students throughout Basic Education, comprised of Early-childhood, Elementary, and High School. The theoretical perspective of the BNCC document focuses on the concept of culture as a social practice, which expresses meanings through language, rather than presenting intrinsic meanings, as for example, artistic events (Moreira & Candau, 2006). This view considers the curriculum as a set of practices that enable the production, circulation, and consumption of meaning in the social sphere and that extensively contribute to the construction of social and cultural identities. This is an important perspective in the discussions regarding curriculum in Brazil, given the diversity of regional cultures and different dialects. As stated by Schmidt et al. (2001),

> our hypothesis is that national culture has an impact on curriculum. We believe it also has an impact on learning. Apart from how culture has an impact

on curriculum and learning separately, culture also has an impact on the relationship between the two. (p. 10)

BNCC includes such considerations. For early-childhood education, it recommends that the school offers a comfortable environment, in which care and social interaction promote socialization, the establishment of affective bonds and trust, as well as activities that promote learning and development. It is essential to create situations in which play, in its many forms, becomes a context that leads to knowledge of one's self, of others, and the world, while taking into account the culture of each community. This should be achieved through friendly interactions, in which attention to one's self and others, as well as attitudes of inquiry, questioning, investigation, and awe are fostered.

In the early years of elementary school, in line with early-childhood education, the guidelines recommend comprehensive care to children, to enable their socialization, while taking care of literacy and the introduction to content systematized by different areas of knowledge. This should be done through activities, which are playful activities, such as play and games; artistic, such as drawing and singing; and scientific, such as the exploration and comprehension of natural and social phenomena. That is why the curricular guidelines for this phase need to integrate many areas of knowledge while focusing on actions that promote literacy. The document contemplates students' learning rights, among which are rights regarding the learning of mathematics in the early years of elementary school and during the literacy cycle. In the Brazilian curriculum, the initial three years of schooling (6 to 8 years of age) are considered the literacy cycle. Mathematical literacy has intended that children have access to and can expand and deepen the mathematical knowledge built on their experiences. It is a process that aims to capitalize on intuitive ideas of children present in mathematical experiences socially and culturally. The appropriation of a language of mathematics contributes to the intellectual development of children.

In the final years of elementary school, due to the change in students' interests as they get older, the playful dimension of the pedagogical practices acquires new characteristics. Such changes must be taken into account in all curricular components, which must also consider the essential continuation of students' social and emotional development. At this stage, new disciplines are introduced, taught by several teachers, which require the commitment to the literacy process in all its aspects (artistic, scientific, humanistic, literary, and mathematical). Therefore, a consistent interdisciplinary articulation is necessary, considering the convergence of relevant themes to different areas of knowledge; literature, history, geography, science, and various disciplines may require distinct mathematical resources in different contexts.

Throughout high school, given the even greater number of disciplines, interdisciplinary articulation is also important within each area of knowledge or among areas. Under this light, the BNCC was organized in four areas of knowledge: Languages, Mathematics, Humanities, and Natural Sciences. This organization aims to overcome the fragmentation in the approach to academic knowledge through the integration and contextualization of such knowledge, respecting the specificities of disciplines that comprise the different areas. In all areas, the learning objectives for the distinct stages of basic education are proposed according to the characteristics of students at each stage, their experiences, and the context of their participation in society. Each discipline defines strands along which the learning objectives are organized. The aim is to articulate disciplines within an area of knowledge, throughout the schooling stages during which content is presented.

Besides these strands, the integration of the disciplines within the same area of knowledge, and among different areas, is established by integrating themes. Such themes comprise issues that inter-connect the experiences and actions of individuals, thus affecting the process of construction of their identity, their interactions with other individuals, and how such interactions shape their ethical and critical disposition towards the world (MEC, 2015). Besides the cognitive dimension, these themes, therefore, contemplate students' political, ethical, and aesthetical education. To address these dimensions, the integrating themes must pervade the learning objectives of the many disciplines at different stages of students' basic education. Such themes include consumer and financial education, ethics, human rights and citizenship, sustainability, digital technologies, and African and indigenous cultures. This curricular structure affects the mathematics curriculum that is discussed in the next section.

The Foundations of National Curriculum Guidelines for Teaching Mathematics

The basic underlying principle of the current Brazilian curriculum is that the objective of mathematics instruction is to achieve a comprehensive understanding of the world and social practices, which enables individuals to enter the world of work, supported by the ability to reason and with the confidence to deal with problems and challenges of different origins. Therefore, it is essential that education is contextualized and interdisciplinary, but at the same time, fosters the development of abilities to abstract, understand what can be generalized to other contexts, and use the imagination.

Basic education curricula include the following guidelines: the dissemination of basic values of social interest, of the rights and duties of citizens, and of respect for the common good and democratic order. Considering these premises, curriculum guidelines for mathematics in basic education are organized in five strands according to learning objectives: Geometry, Quantities and Measurements, Statistics and Probability, Numbers and Operations, and Algebra and Functions. Each of the strands merits different emphasis depending on the year, seeking to ensure that students' proficiency in mathematics becomes increasingly sophisticated over the years of schooling. The goals for each strand, within one school year, are selected and planned to establish connections between knowledge of different strands and different disciplines so students can discover a wealth of knowledge (MEC, 2015).

In the first three years of elementary school, dedicated to literacy, children are expected to further improve their location systems and the ability to describe space, which is complemented by experiments with different magnitudes that surround us, and allow successive approximations to geometry. Through the initial contact with probability and statistics, students begin to comprehend the uncertainty of mathematics as an object of study, and for example, its role in understanding social issues for which answers are not always unique and conclusive. Within the strand of Numbers and Operations, students are expected to gain autonomy in numerical thinking, without the constraints of conventions and unnecessary formalism. Thus, the objective is that students have access to and understand the order of magnitude of numbers, as that is where the strength of the understanding of the decimal system lies. The aim is that students understand and perform operations using strategies that make sense for them, and that such strategies are evaluated, compared, and improved.

Algebra is associated with the ability to identify attributes and rules for forming sequences, which is the first evidence of organization of thought. In addition, the first indicator of the idea of function is the ability to recognize changes and relationships.

In the fourth and fifth years of elementary school, the systematization of Geometry begins with students understanding the characteristics and properties of plane and spatial figures. Regarding Quantities and Measurements, the knowledge of the International System of Units strengthens and gives structure to the concept of quantities, enabling students to develop autonomy to deal consciously and critically with day-to-day commercial and financial issues. In the area of Statistics and Probability, understanding the randomness and uncertainty of different situations enables a better understanding of social issues, which are useful for the construction of values, and a more critical analysis of the information pervading everyday life. To learn all this, it is essential to expand students' knowledge of natural numbers and operations, as well as provide initial contact with the positive rational numbers. Such knowledge should always arise from contextualized situations and problems, and should be organized so it can be decontextualized from specific applications and applied to new problem-solving situations.

Thus, school mathematics is constructed following the development of students through successive discoveries of possibilities and concepts. In this way, students start making sense of solutions to new problems. A prime example is the field of numbers, which is expanded by students' realization that natural and positive rational numbers are not sufficient to explain new situations. Negative numbers and numeric sets, integers and rational numbers, as well as real numbers are discovered at this stage. Similarly, through the other strands, students may come to realize that new objects of knowledge are necessary to meet social and scientific demands, such as composite numbers, more precise localizations through the Cartesian plane, and understand how to deal with statistical data (e.g., students become increasingly proficient in reading and interpreting data to infer results). It is at this stage that students gain depth of knowledge about Algebra and Functions, especially regarding their understanding of variability and dependence, which helps enhance their logical reasoning and expand their problem-solving abilities.

We consider that learning mathematics in high school would need to prioritize concepts and procedures that enable students to establish connections between both distinct mathematical ideas and other areas of knowledge, with significant emphasis to their social applications. The study of functions should prioritize aspects related to the variation between quantities, allowing students to develop functional thinking effectively, thus substituting for skills related to simple symbolic-algebraic manipulation.

Working with quantities and measures is intended to promote the integration and coordination between different fields of knowledge. The exploration of geometry may constitute a stimulus for the students to include demonstrations of geometric figures, promoting the expansion and consolidation of previously learned concepts. For example, geometric exploration requires the use of measures, which leads to convergence with other disciplines, such as physics and chemistry.

In high school, the study of numbers should favor the perception of groups in different number sets. The comprehension of the limitations of irrational numbers should promote students' understanding of real numbers due to the need to increase numerical knowledge.

The document suggests enriching the work with mathematics in high school through challenges based on the use of technological resources as tools to assist students learning and completing projects. The objectives of mathematics in high school are to apply mathematical knowledge to different situations, in order to understand other sciences and consolidate a comprehensive scientific education. High school mathematics also entails expressing oneself orally, in writing, and graphically, valuing language accuracy, the communication of ideas and mathematical reasoning; understanding mathematics as a science, with its own language and logical-deductive structure; and establishing relationships between mathematical concepts within the same strand and among the different areas. In addition, it is recommended that concepts from mathematics and from other areas of knowledge are connected, leading to critical analyses of the use of mathematics in different social practices and to understand natural phenomena, in order to function and intervene in society (MEC, 2015).

The Influence of Research in Mathematics Education in the National Curriculum

In recent years, Brazilian curriculum developers have engaged in dialogue with researchers in mathematics education. Two great areas of research in mathematics education significantly influenced the National Curriculum Parameters issued in 1997: studies on Ethnomathematics and the International Problem Solving Movement.

Ethnomathematics as a research field has been driven by studies conducted by Ubiratan D'Ambrosio (1990) in his Ethnomathematics Program: art or technique to explain and comprehend (*Programa de Etnomatemática: arte ou técnica de explicar e conhecer*) and by the studies of Terezinha Nunes (Carraher), David Carraher, and Ana Lucia Schiliemann (1988). Such studies showed that the mathematics children were learning in school made little sense in their social practices; children were able to successfully use mathematics in their business dealings outside the classroom, but were not able to perform the same traditional calculations in school. The studies in this field, expanded by researchers who examined other social practices, showed that different social groups have distinct forms of mathematization that must be recognized by the school, which must legitimize these different mathematical practices as well as grant students access to the (traditional) historically produced mathematics. This anthropological and social perspective of recognition of school mathematics as one of the disciplines of the mathematics curriculum interconnects with another field of research in mathematics education, problem-solving, which was heavily influenced by American documents, such as *An Agenda for Action* (National Council of Teachers of Mathematics, 1980) and *Standards* from the National Council of Teachers of Mathematics in the United States of America (1989, 2000). These documents highlighted mathematics as part of cultural heritage and as important to the scientific and technical community. The vision of school mathematics in these documents was based on students learning mathematics with understanding. These ideas influenced this Brazilian scenario and in the 1990s several research groups were created in Brazil, such as *Grupo de estudos sobre resolução de problemas* (Study Group on problem-solving) of Universidade Estadual Paulista "Júlio de Mesquita Filho"—UNESP/Rio Claro and *CEM—Centro de Educação Matemática* (Mathematics Education Center) at Universidade de São Paulo, that conducted research in mathematics education. The research on problem-solving made it possible for educators to engage their students in actual problem situations, so they could raise hypotheses, develop resolution strategies, try them, reflect on the results, create new problems, and model them. In this sense, the National Curriculum Parameters valued teaching strategies involving Ethnomathematics, the use of the history of mathematics for teaching, technology, mathematical modeling, problem solving, games, and manipulative materials.

The BNCC includes aspects related to problem solving under a perspective of mathematics as research. The goal for students is to assign meaning to mathematical knowledge to solve problems arising from their social practices. It is important that the relationship of students with math induces self-confidence, through participation in attractive and challenging experiences. The document recommends favoring interdisciplinary work connecting mathematics to other fields of knowledge, as well as within mathematics itself. However, there has been criticism in this respect, as the introductory part of the BNCC does not seem to be aligned with the objectives and guidelines proposed in the document. As noted by Passos (2016):

> The apparent emphasis on interdisciplinary work presented in the introduction of the document is not corroborated by the learning objectives, which configure a sort of "self-centered mathematics." It seems strange that a document that seems to value interdisciplinarity does not refer clearly and objectively to the trends in mathematics education, discussed and surveyed both in Brazil and abroad, such as: Ethnomathematics, IT in Mathematics Education, Mathematical Modeling, Critical Mathematics Education, and the Cultural Historical Approach for teaching mathematics. The manner through which learning objectives are presented refers to a standard of large-scale external evaluations. The document does not bring the work with interdisciplinarity in the list of the learning objectives. Too, isn't considered the peculiarities of each region and each school and this is very important for the interdisciplinarity.

> Regarding geometry, for example, it incorporates the results of research advocating learning based on intuition and experimentation. The experiments are based on manipulations, imagery production, representation, as well as description and recognition properties (Nacarato & Passos, 2003). However, there are few references to the connection between plane and solid geometry.

Metric is valued at the expense of conceptual training in geometry. The construction of geometric theory made possible by discussion, evidence and demonstration (Nasser & Tinoco, 2001; Pais, 1996) is an afterthought, restricted to a project towards the end of elementary school. (p. 23)

The strand of Statistics and Probability incorporates the research of Lopes (1998, 2003) and Souza (2007, 2013) to stress the importance of students recognizing uncertainty as an object of study for mathematics, and understanding the meaning of words such as chance, randomness, sampling, and variability. Moreover, to indicate the understanding of important concepts, such as distribution, position measures, and dispersion, students conduct statistical research projects covering all stages of data collection, organization, interpretation, and analysis.

Recently, research regarding Algebra and Functions has examined algebraic initiation, the knowledge of students regarding elementary algebra, the meaning of equations, the equals sign, and teachers' knowledge of functions (Ribeiro & Cury, 2015; Zuffi, 1999). The orientation of such research subsidizes an approach to algebraic initiation based on natural language from the students' everyday life, with an appropriation of algebraic language through the course of schooling. The importance of understanding algebraic symbols, thus avoiding mechanical memorization of rules and formulas, is also considered. As stated by MEC (2015): "The study of functions should prioritize aspects related to the variation between quantities, allowing the student to develop functional thinking effectively, replacing the skills related to simple symbolic-algebraic manipulation, usually favored in school" (p. 140).

THEORETICAL GUIDELINES OF MATHEMATICS EDUCATION THAT HAVE INFLUENCED THE DEVELOPMENT OF THE CURRICULUM

The BNCC does not detail the theoretical framework, which supports the concepts regarding mathematics teaching and learning, and how the areas of knowledge included in the curriculum are linked. However, it is possible to infer some theoretical premises which support the learning and teaching objectives for Basic Education.

One objective is the focus on contextualization, that is, the possibility of thinking about problem situations rooted in the students' day-to-day social practices. The idea is that students have knowledge arising from their social practices, though not yet formal knowledge, which contributes to reflection and mobilization in problem-solving situations.

Another focus is on different languages used to present mathematical ideas. Thus, the student begins with spontaneous writing and gradually takes ownership of historically produced mathematical language. The role of the school is to afford the possibility of expanding the linguistic repertoire of young students throughout the schooling period.

Another focus is on problem solving and the ability to do math, to experience situations, propose solutions, and solve problems. This idea is mentioned in the curriculum document MEC (2015): "Thus, learning mathematics requires the exploration of three distinct and sequenced stages. In the beginning, students must do math. After that, they must develop *individual representation registries*, and finally take *ownership of formal records*" (p. 117).

We believe this concept of knowledge acquired in a linear, orderly, and dissociated manner, without articulation, represents an oversimplification in terms of theoretical knowledge regarding mathematics education. We advocate, based on Vygotsky's theoretical perspective, that learning does not occur in a linear fashion and in an orderly manner because thought is built gradually in a historical and essentially social environment. For Vygotsky (1988), social interaction plays a key role in cognitive development and cultural development of a subject, because it enables the generation of new experiences and knowledge. To assume this perspective for mathematical education is to enable students learning as a social experience.

Finally, there is the focus on interdisciplinarity encompassing the link between fields of knowledge, modeling mathematical problems in real situations, and resolving problems within mathematics itself. Although this concept is developed in the first part of the document, there are few connections to other fields of knowledge within the objectives of mathematics teaching. Thus, we believe that the discussions in the curriculum documents currently under review in Brazil fail to include theoretical advances that have been achieved in the field of mathematics education.

Engaging Teachers in Discussion About the Documents

The debate regarding the BNCC is still restricted to academia, and few schools have systematically discussed it or reflected on it. Even though the document can be obtained from the Ministry of Education, the school community (teachers, parents, and students) has shown little interest in becoming familiar with its content.

Teachers only take part in curriculum debates where they must read, discuss, and comment on the document when convened by school officials. Parents show little interest in curriculum guidelines; rather, they concentrate on the evaluation results achieved by their children. Although students regularly question the content to be learned, especially regarding mathematics, they are not interested in discussing curricular issues in the final years of elementary school and during high school, unless some sort of class project is proposed by the school (Carvalho, 2000).

High school students are concerned with the entrance examinations to be accepted by universities, as these always require thorough mathematical knowledge that is not always consistent with what they had the opportunity to study at school. This selection process to universities relies on tests that do not value the quality of doing mathematics, the value is only in the response and not in the process. Moreover, the designers of these tests rarely take into account the recommendations made by the official curriculum. Each university develops its own mathematical content list that will be contained in their tests.

Problems Faced During Curriculum Implementation

The greatest challenge facing curricular reform in Brazil undoubtedly is teacher training. The national curriculum documents are drawn up by experts who, in most cases, are academics and university professors, who may know little about the reality of Brazilian classrooms. As a result, teachers do not feel included in the proposals and require a long period of discussion, adjustment, and understanding before deciding to implement them. In fact, implementation is driven by the textbooks that adopt the new curriculum as their frame of reference. When evaluated and acquired by the schools, these books become the curriculum reference for teachers. Teachers also appropriate new content because they know that external evaluations will include such content, leading to attempts to "drill" students so they can perform well.

Curriculum documents are not always clear to teachers and require the subsequent publication of teaching guidelines and books, as well as continuing teacher training if implementation is to occur. Brazilian researchers in mathematics education have pointed this out while examining issues related to mathematics curriculum development in basic education, and aspects related to curriculum policy and practice (Costa, 2011; Godoy, 2011). Some research groups in mathematics education, listed on the roster of *Conselho Nacional de Desenvolvimento Científico e Tecnológico* [National Council for Scientific and Technological Development] (CNPQ) have been conducting studies and discussing specific aspects of the mathematics curriculum. Among such groups the following are worth mentioning: *Estudos Curriculares em Educação matemática* [Studies on Mathematics Education Curriculum] in the state of Rio Grande do Sul; *Grupo de Pesquisa Currículo e Educação Matemática* [Research Group on Curriculum and Mathematics

Education] in the state of Mato Grosso do Sul; *História da Educação Matemática: aspectos históricos, curriculares e culturais* [History of Mathematics Education: historical, curricular and cultural aspects] in the state of Minas Gerais; *Formação de Professores: Currículo, História, Linguagem e Desenvolvimento Profissional* [Teacher Training: Curriculum, History, Language and Professional Development] in the state of São Paulo; and *Grupo de Estudos e Pesquisas em Educação Matemática e Educação* [Study and Research Group in Mathematics Education and Education] in the state of São Paulo.

Over the 10-year period since its publication, experience with the National Curricular Parameters has shown that teachers have taken ownership of many of the concepts, teaching guidelines, or proposed assessment schemes in the document. Once such a document is published, many things are set in motion. The continuing teacher education and initial training courses for teachers begin to discuss the documents, textbooks are redesigned, research studies in the area start to be developed, partnership projects between universities and schools are encouraged, and teachers contemplate the changes offered by the new curriculum. These efforts require great financial investments by municipal, state, and federal governments. Once again, the challenges are numerous in a country of continental proportions, where some regions can only be accessed after long boat journeys.

FINAL THOUGHTS

In Brazil, public policy pressure systematically modifies the curriculum guidelines for teaching and learning mathematics. There have been changes in curriculum conceptions and methodological recommendations. However, the mathematical content requirements for each educational level have not changed significantly in the past four decades. We believe we should move away from a long linear list of mathematical content recommendations for each school year, as it compromises the significant development of mathematical thinking and the comprehension of multiple mathematical concepts.

The lack of investment in the participation of teachers in the development of curriculum guidelines may be a complicating factor in the curriculum implementation process. As D'Ambrosio and Lopes (2014) pointed out, "curriculum guidelines established by educational authorities or school systems sometimes disqualify teachers, as they do not include them in the preparation of such guidelines" (p. 79). In addition, sometimes experts develop curriculum guidelines without being aware of the actual situation in many school classrooms, which are extremely socially, culturally, and economically diverse throughout the country. Moreover, there does not seem to be a consensus within mathematics education regarding the theoretical framework, beliefs, and premises related to teaching and learning mathematics that directly influence the curriculum proposed. Each curriculum proposal is laden with the ideas, directives, and premises of those responsible for writing it.

Curriculum thinkers and developers need to resize their recommendations in relation to the learning of children and youngsters, and this requires creative insubordination. For D'Ambrosio and Lopes (2014), creative insubordination refers to opposition, and generally means challenging established authority when it does not seek the common good. It means to be aware of when, how, and why one should take action against established guidelines or procedures. The current development of a mathematics curriculum requires reacting to the *status quo*; it is necessary to assume that, in view of the current scenario in society and its technological development, there is no longer any justification for the study of certain mathematical content in basic education. "Creative insubordination attitudes regarding the curriculum will generate a rupture with the idea of an unreal, artificial, unified, standard and linear curriculum" (D'Ambrosio & Lopes, 2014, p. 80).

In addition, it is necessary to connect the production of curriculum documents to effective public policies, which support study and teaching conditions, thus enabling the implementation of the new guidelines. One must consider the need for sustainability of curriculum documents and public policy, as it has become a tradition in states and cities in Brazil that a change of government brings about changes both in the curriculum and in educational management.

ACKNOWLEDGMENT

This chapter was written in honor of Beatriz Silva D'Ambrosio.

REFERENCES

Barretto, E. S. S. (2012). Políticas de currículo e avaliação e políticas docentes [Curriculum and evaluation policies and teacher policies]. *Cadernos de Pesquisa*. São Paulo, *42*(147), 1–16.

Carraher, T., Carraher, D., & Schliemann, A. (1988). *Na vida dez, na escola zero* [In life ten, at school zero]. São Paulo, Brazil: Cortez.

Carvalho, M. E. P. de. (2000). *Family-school relations: A critique of parental involvement in schooling*. Mahwah, NJ: Lawrence Erlbaum Associates.

Costa, J. C. O. (2011). *O currículo de matemática no ensino médio do Brasil e a diversidade de percursos formativos* [The mathematics curriculum in high school in Brazil and the diversity of training courses] (Doctoral dissertation). Faculdade de Educação, Universidade de São Paulo, São Paulo.

D'Ambrosio, B. S. (1987). *The dynamics and consequences of the modern mathematics reform movement for Brazilian Mathematics Education* (Doctoral dissertation). School of Education, Indiana University.

D'Ambrosio, B. S., & Lopes, C. E. (2014). *Trajetórias profissionais de educadoras matemáticas* [Professional trajectories of mathematical educators]. Campinas, São Paulo, Brazil: Mercado de Letras.

D'Ambrosio, U. (1990). *Etnomatemática* [Ethnomathematics]. São Paulo, Brazil: Ática.

Garnica, A. V. M. (2014). Brief considerations on educational directives and public policies in Brazil regarding mathematics education. In Y. Li & G. Lappan (Eds.), *Mathematics curriculum in school education: Advances in Mathematics Education* (pp. 143–156). Dordrecht, the Netherlands: Springer.

Godoy, E. V. (2011). *Currículo, cultura e educação matemática: uma aproximação possivel?* [Curriculum, culture and mathematical education: a possible approach?] (Doctoral dissertation). Faculdade de Educação, Universidade de São Paulo, São Paulo, Brazil.

Instituto Nacional de Estudos e Pesquisas Educacionais Anísio Teixeira (INEP). (2014). *Censo Escolar da Educação Básica 2013: Resumo técnico* [School Census of Basic Education 2013: Technical summary]. Brasília: Author.

Lopes, C. E. (1998). *A probabilidade e a estatística no ensino fundamental: uma análise curricular* [Probability and statistics in elementary school: A curricular analysis] (Master's thesis). Faculdade de Educação, Universidade Estadual de Campinas, Campinas, São Paulo, Brazil.

Lopes, C. E. (2003). *O conhecimento profissional dos professores e suas relações com Estatística e probabilidade na educação infantil* [The professional knowledge of teachers and their relationships with statistics and probability in childhood education] (Doctoral dissertation). Faculdade de Educação, Universidade Estadual de Campinas, Campinas, São Paulo, Brazil.

Lopes, C. E., Santos, M. C., Gravina, M. A., & Carvalho, P. C. P. (2006). Conhecimentos de Matemática [Mathematical Knowledge]. In Brasil, *Orientações curriculares para o ensino médio: ciências da natureza, matemática e suas tecnologias* [Curricular guidelines for high school: Natural sciences, mathematics and their Technologies] (pp. 69–98). Brasília: Ministério da Educação, Secretaria de Educação Básica.

Miguel, A., Garnica, A. V., Igliori, S., & D'Ambrosio, U. (2004). A educação matemática: Breve histórico, ações implementadas e questões sobre sua disciplinarização [Mathematics education: Brief history, actions implemented and questions about its disciplinarization]. *Revista Brasileira de Educação*. Set/Out/Nov/Dez, no. 27. p. 70–93. Retrieved from http://www.scielo.br/pdf/rbedu/n27/n27a05.pdf

Ministério da Educação e do Desporto do Brasil. (MEC). (1997). *Parâmetros Curriculares Nacionais* [National Curricular Guidelines]. Secretaria de Educação Fundamental. Brasília: MEC/SEF.

Ministério da Educação e do Desporto do Brasil. (MEC). (1998). *Referencial curricular nacional para a educação infantil* [National curriculum guidelines for childhood education]. Secretaria de Educação Fundamental. Brasília: MEC/SEF.
Ministério da Educação e do Desporto do Brasil. (MEC). (2000). *Parâmetros Curriculares Nacionais para o Ensino Médio* [National curriculum guidelines for high school]. Secretaria de Educação Fundamental. Brasília: MEC/SEF.
Ministério da Educação e do Desporto do Brasil. (MEC). (2013). *Diretrizes Curriculares Nacionais Gerais da Educação Básica* [General National Curricular Guidelines for Basic Education]. Secretaria de Educação Básica. Diretoria de Currículos e Educação Integral. Brasília: MEC/SEB.
Ministério da Educação e do Desporto do Brasil. (MEC). (2015). *Base Nacional Comum Curricular (BNCC)* [National Common Curricular Base]. Secretaria de Educação Básica. Brasília: MEC/SEB.
Moreira, A. F., & Candau, V. M. (2006). Currículo, conhecimento e cultura [Curriculum, knowledge and culture]. In Brasil, *Indagações sobre currículo* [Questions about curriculum]. Brasília: Ministério da Educação.
Nacarato, A. M., & Passos, C. L. B. (2003). *A Geometria nas séries iniciais: Uma análise sob a perspectiva da prática pedagógica e da formação de professores* [Geometry in the elementary school: An analysis from the perspective of pedagogical practice and teacher education]. São Carlos, São Paulo, Brazil: EdUFcar.
Nasser, L., & Tinoco, L. A. A. (2001). *Argumentações e provas no ensino de matemática* [Arguments and proofs in mathematics teaching]. Rio de Janeiro, Brazil: Projeto Fundão, UFRJ.
National Council of Teachers of Mathematics. (1980). *An agenda for action.* Reston, VA: Author.
National Council of Teachers of Mathematics. (1989). *Curriculum and evaluation standards for school mathematics.* Reston, VA: Author.
National Council of Teachers of Mathematics. (2000). *Principles and standards for school mathematics.* Reston, VA: Author.
Pais, L. C. (1996, julho/dezembro). Intuição, experiência e teoria geométrica [Intuition, experience and geometric theory]. *Zetetiké, 4*(6), 65–74.
Passos, C. L. B. (2016). *Parecer sobre documento da base nacional comum curricular. Matemática* [Opinion on document of the common national curricular basis. Mathematics]. *Ensino Fundamental.* Unpublished manuscript.
Pires, C. M. C. (2008). Educação Matemática e sua Influência no Processo de Organização e Desenvolvimento Curricular no Brasil [Mathematical education and its influence on the process of organization and curricular development in Brazil]. Revista Eletrônica. *Bolema, 29,* 13–42.
Ribeiro, A. J., & Cury, H. N. (2015). *Álgebra para a formação do professor: Explorando os conceitos de equação e de função* [Algebra for the education of the teacher: Exploring the concepts of equation and function]. Belo Horizonte, Minas Gerais, Brazil: Autêntica.
Schmidt, W. H., Schmidt, W. H., McKnight, C. C., Houang, R. T., Wang, H., Wiley, D. E., . . . & Wolfe, R. G. (2001). *Why schools matter: A cross-national comparison of curriculum and learning.* San Francisco, CA: Josey-Bass.
Souza, A. C. (2007). *A Educação Estatística na infância* [Statistical education in childhood] (Master's thesis). Universidade Cruzeiro do Sul, São Paulo, Brazil.

Souza, A. C. (2013). *O desenvolvimento profissional de educadoras da infância: Uma aproximação à Educação Estatística* [The professional development of childhood educators: An approach to statistical education] (Doctoral dissertation). Universidade Cruzeiro do Sul, São Paulo, Brazil.

Vygostky, L. S. (1988). *A Formação Social da Mente* [The social formation of mind]. São Paulo, Brazil: Martins Fontes.

Zuffi, E. (1999). *O tema "funções" e a linguagem matemática de professores do ensino médio: por uma aprendizagem de significados* [The theme "functions" and the mathematical language of high school teachers: through a learning of meanings] (Doctoral dissertation). Faculdade de Educação, Universidade de São Paulo, São Paulo, Brazil.

CHAPTER 9

THE KOREAN MATHEMATICS CURRICULUM
Characteristics and Challenges

Kyeong-Hwa Lee
Seoul National University

JinHyeong Park
Myongji University

Na-Young Ku
Anyang High School

Curriculum is a term with various meanings in mathematics education, from national curriculum to hidden curriculum (Remillard & Heck, 2014). In the Korean context, curriculum usually refers to the national curriculum (Kang, 2013; Paik, 2004; Park, 1997). A uniform mathematics curriculum at the national level has been administered in Korea since the establishment of the Republic of Korea in 1948. The first mathematics curriculum was in the form of a syllabus that contained a list of what to teach for each grade but without any detailed background information. However, the Korean War occurred during the first curriculum implementation. In the turbulence of the post-war period, the Korean government implemented strong reforms to reconstruct the nation.

This first mathematics curriculum reform in 1955 was strongly influenced by Deweyan pragmatism and emphasized mathematics in everyday life. Similarly, major educational reform ideas, such as enhancing *problem solving, mathematical communication, mathematical reasoning, self-regulated learning,* and *competency-focused approach,* that had been established in advanced countries were accepted and disseminated by the group of researchers authorized by the government. Moreover, based on the aspiration to rebuild the country, school mathematics had an indispensable position among school subjects in education. Hence, major suggestions for mathematics education reform were actively reflected in the curriculum revisions. The fifth mathematics curriculum in the late 1990s emphasized problem solving and the seventh mathematics curriculum in 1997 claimed to advocate a student-centered curriculum. The revised mathematics curriculum in 2007 added teaching and assessing mathematical processes, such as mathematical reasoning, connections within and out of mathematics, and mathematical communication, to the learning of mathematical content.

The Korean mathematics curriculum has been revised periodically in ways that reflect social and educational needs that have emerged inside and outside of mathematics education and that echo voices from leading international research groups. Reinterpretation and adaptation of major reform ideas, such as promoting problem solving, reasoning, communication, and connections, were done in the process of making revisions. While major ideas have been adapted from international research trends and reflected in the written curriculum, researchers and administrators have tried not to lose the traditional Korean mathematics education perspective that emphasizes basic knowledge and basic skills in the teaching and learning of mathematics. In addition, there have been dynamic changes and ongoing issues in Korean educational contexts, such as an increasing number of students from immigrant families and from North Korea, students' negative attitudes toward mathematics and mathematics learning, and recognizing and developing gifted students in their early years. Thus, the Korean mathematics curriculum is always changing so its characteristics can be kept current.

Although the Korean mathematics curriculum has tried to integrate trends from other nations, it has its own characteristics and faces its own challenges. The characteristics and challenges that the Korean mathematics curriculum faces need to be understood from a historical point of view because the curriculum has changed a great deal based on the needs and the hopes that come from this background. In the following sections, we give a picture of a slice of the Korean curriculum, focusing on the recently revised curriculum that reveals the contemporary visions and purposes of Korean mathematics educators.

INTENDED MATHEMATICS CURRICULUM

This section introduces two major revisions of the Korean mathematics curriculum: the seventh mathematics curriculum in 1997 and the revised mathematics curriculum in 2015. The seventh mathematics curriculum pursued *student-centered* or *learner-centered* curriculum and this revision was regarded as groundbreaking compared to previous curricula (Lew, 1999). The revised mathematics curriculum in 2015, a recently reformed mathematics curriculum, focused on promoting students' mathematical competencies. The review of these two main reforms of the Korean mathematics curriculum will show the contemporary direction of the Korean intended mathematics curriculum.

TRANSITION FROM CONFUCIAN PEDAGOGY TO COMPETENCE-BASED EDUCATION

It has been noted that the Korean curriculum is historically based on Confucianism (Kim, 2007; So, Kim, & Lee, 2012). In Confucianism, learning was a process of reading, understanding, and deliberating (Yao, 2000), and Korean traditional education was performed in this way. So, Kim, and Lee (2012) pointed out that "Confucian pedagogy was called *yuhak*. The main educational method of *yuhak* was interpreting and memorizing Chinese scriptures" (p. 798).

Although the focus of Korean traditional education moved to Western disciplines in the early 20th century, "Koreans experienced severe conflict and confusion due to the pro-democracy movement that aimed to dismantle the military dictatorship and the corresponding oppressive measures of the military government" (So, Kim, & Lee, 2012, p. 799) until establishment of a civilian government in the 1990s. The Korean curriculum underwent big changes in the seventh curriculum revision in 1997, stimulated by criticisms on the weak points in the existing curriculum. For example, Lew (1999) pointed out that "compared to that of 30 years ago, the current [mathematics] curriculum shows few changes either in content or in methods of teaching and evaluation. It is skill orientated, relying on the expository method to transfer fragmentary pieces of knowledge" (p. 218). The seventh curriculum revision pursued the *learner-centered* curriculum (Shin et al., 2005). The term *learner-centered* curriculum highlights mathematics learning through students' autonomy and creativity rather than teacher-directed learning by memorization and repetition. According to Lew (1999), the seventh curriculum revision adopted a "constructivist view emphasizing students' own learning processes and their own ways of appreciating the values of knowing" (p. 218). To be specific, the seventh curriculum emphasized encouraging students to set up their own learning plans and goals, having them engage in inquiries, and assessing their explorations (Kheel, 2002). In addition, students were given opportunities to select subjects based on their own ability, interest, aptitude, and career plans, though the types of opportunities available were limited (Shin et al., 2005).

Two decades of exploration of various reform ideas that occurred over several revisions after the seventh curriculum revision resulted in a Korean mathematics curriculum that came to have innovative characteristics in terms of teaching philosophy, learning principles, methods, evaluation, and so on. For example, cultivation of mathematical creativity was adopted as a major aim of the mathematics curriculum in the revised mathematics curriculum in 2007. Moreover, teaching and learning mathematical processes and process-focused assessments were added and emphasized in the revised mathematics curriculum in 2007 and 2011. However, in 2015, other criticisms, conflicts, and confusion rose to the surface of mathematics education when the mathematics curriculum revision was conducted. The essential issue was related to different stances on the appropriateness of the amount of learning content contained in the curriculum; nonetheless, the revision team focused on how to realize *competence-based education* in the mathematics curriculum advocated by the government. The Korean government required all subjects, including mathematics, to promote students' competences rather than fragmented knowledge, accepting the meaning of competences from the DeSeCo (Definition and Selection of Competences) project by the Organization for Economic Cooperation and Development (OECD; Lee, Min, Jeon, Kim, & Kim, 2008). Competence was interpreted as an ability to process and apply a vast store of knowledge and information to create new knowledge (So, 2006). The mathematics curriculum was also designed to build key competences as well as learning mathematical content (Kim et al., 2009). Specifically, the revised mathematics curriculum in 2015 focused on six key competences: problem solving, communication, reasoning, creativity and convergence, attitude and practice, and data processing (Korean Ministry of Education [KMOE], 2015a).

Common Plus Elective Curriculum

The revised Korean mathematics curriculum is comprised of 10 years of common curriculum from elementary school to the first year of high school and two years of elective curriculum. (See Figure 9.1 for details of the mathematics curriculum in relation to the school system.) In the elective curriculum, students can select mathematics subjects based on their needs and interests. For example, students who want to continue their studies at a university can choose Math II, and either Calculus or Probability and Statistics. Students who are in vocational high schools choose Geometry and either Mathematics in Real Life or Mathematics in Economy.

The document for the revised mathematics curriculum in 2015 includes the overall aim, objectives, content and achievement standards, and general

age				
17		Elective-centered curriculum	Elective mathematics subjects	3rd year	High school
16				2nd year	
15				1st year	
14		The national common basic curriculum	Common mathematics subjects	3rd year	Middle school
13				2nd year	
12				1st year	
11	Compulsory education			6th year	Elementary school
10				5th year	
9				4th year	
8				3rd year	
7				2nd year	
6				1st year	
5		The kindergarten curriculum	Kindergarten		
4					
3					
2					
1					
0					

Figure 9.1 Intended curriculum in relation to the school system.

guidelines for teaching and evaluation (see Figure 9.2.). Both general and specific guidelines that are based on relevant research for teaching, learning, and evaluating mathematics are given in the document.

The overall aim relates the general human characteristics, such as openness, creativity, and altruism, that mathematics education aims to nurture. Objectives are more specific goals to be pursued by mathematics education, such as obtaining and understanding fundamental knowledge and skills in mathematics, learning how to think and communicate mathematically in order to investigate diverse phenomena, and solving problems mathematically. Content and achievement standards define the range and the scope of the content to be learned, with specific guidelines for teaching and learning it. For example, the concept of function in middle school (from seventh to ninth grade) is presented as content with a description whereby teachers can recognize what key concepts are to be taught, what generalized knowledge is aimed for, what learning elements are to be included, and what skills are to be developed (see Table 9.1). The linear function and quadratic function are first introduced to middle school students and higher polynomial functions and transcendental functions are introduced in the high school mathematics curriculum.

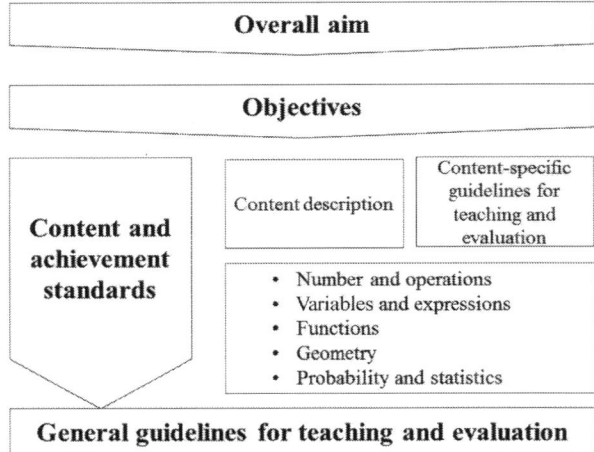

Figure 9.2 Organization of the document for the revised mathematics curriculum.

TABLE 9.1 Content Standards of Functions for Middle School				
Strand	Key Concepts	Generalized Knowledge	Learning Elements	Skills
Function	Function Graph	A function represents a relationship between two changing quantities. A function involves a relationship that can be a correspondence or a dependency between two quantities. A graph is a tool for visualizing functions.	Plane coordinates Linear functions and graphs The relationship between a linear function and a linear equation The quadratic function and its graph	To understand To interpret To represent To draw graphs To solve problems To apply To explore

Content-specific guidelines include teaching and learning strategies as well as assessment methods that relate relevant research with the main learning elements. For example, many researchers have encouraged the introduction of real-world examples of plane coordinates for effective teaching and learning.

POTENTIALLY IMPLEMENTED CURRICULUM

Public education (i.e., the idea that education should be provided and useful to all) is a deep-rooted perspective in Korean mathematics education. There has been a great deal of discussion about how the government and social institutions, such as schools, are jointly responsible for providing equal chances among children regardless of their home backgrounds (Sung, 2007): the mathematics curriculum should be developed and implemented for all students and not just a certain segment of the population. The range and scope of content to be learned has traditionally been controlled by a committee authorized by the government. Textbooks, the potentially implemented curriculum, undergo a strict qualification process by the authorizing committee. The only government-authorized textbook for each grade in elementary school is developed by a group of writers composed of professors and teachers authorized by the KMOE. Although there are several kinds of textbooks developed for secondary schools by different groups consisting of professors and teachers, variation among the textbooks is much lower than that of other countries (e.g., the United States). Textbooks with a limited range and scope of content could be one reason why variation among Korean schools is much lower than that of other countries in international comparative studies, such as Trends in International Mathematics and Science Study (TIMSS) and Programme for International Student Assessment (PISA; OECD, 2002).

The idea of public education means that the responsibilities and ethics of relevant specialists in mathematics education, such as policy makers, administrators, mathematics educators, mathematics textbook writers, and mathematics teachers, are stressed explicitly and implicitly. In particular, textbook writers take responsibility to meet the content standards set in the intended curriculum. Moreover, textbooks should realize what the intended curriculum presents as overall aims and objectives by providing rich opportunities with appropriate tasks and activities for teaching and learning mathematics. The responsibilities and ethics expected for textbook writers are not only about individual matters but also about authorization of the textbooks. In other words, the KMOE sets the criteria for evaluating mathematics textbooks, and mathematics textbooks can be published only if they satisfy the criteria (see Table 9.2).

Textbooks have been carefully analyzed from various points of view in relevant research. For example, Chang, Kim, and Lee (2013) analyzed to what

TABLE 9.2 Criteria for Evaluating Mathematics Textbooks

Category	Criteria	Weight
I. Consistency with the intended curriculum	1. Is it consistent with the characteristics of the intended curriculum?	20
	2. Is it consistent with objectives of the intended curriculum?	
	3. Is it consistent with content and achievement standards of the intended curriculum?	
	4. Is it consistent with guidelines for teaching and evaluation of the intended curriculum?	

(continued)

TABLE 9.2 Criteria for Evaluating Mathematics Textbooks (continued)

Category	Criteria	Weight
II. Selection and organization of content	5. Is it relevant and optimized in terms of content?	25
	6. Is the content organized in a way that students will understand easily?	
	7. Is the content organized considering hierarchy and systemicity?	
	8. Is the content organized such that understanding concepts, principles, and laws through social and natural phenomena and acquiring skills is in accordance with the level of students?	
	9. Is the content organized to support cultivating relevant mathematical competences?	
	10. Are data such as photos, illustrations, graphs, and statistics in harmony with content? Are the references clearly presented?	
III. Accuracy and fairness of content	11. Are mathematical concepts, principles, laws, terms, and symbols described accurately?	25
	12. Are the problems and solutions error-free? Are all the solutions to the problems presented?	
	13. Is it free from slander, distortion, or support related to a particular region, culture, hierarchy, personality, gender, goods, institutions, religion, population, or occupation? Is it written fairly without any of the textbook writers' prejudices?	
	14. When data are presented, are they in compliance with the latest laws and regulations including intellectual property rights and patent rights?	
	15. Is notation such as Hangul, Chinese characters, the Roman alphabet, names of people, places, various terms, statistics, graphs, maps, and weighing units accurate? Does it follow the standards for textbooks faithfully?	
	16. Is it written in proper words without mistakes in spelling, grammar, and vocabulary?	
IV. Teaching and evaluation	17. Does it utilize a variety of teaching methods to cultivate mathematical competences?	30
	18. Does it actively and properly use technology and teaching materials?	
	19. Are problems and tasks consistent with objectives, content and achievement standards, and guidelines for teaching and evaluation?	
Total		100

Source: KMOE (2015b)

degree and in what aspects newly developed textbooks reflected the intended elementary mathematics curriculum. They found that new emphases highlighted in the intended curriculum, such as mathematical processes, were properly included in the newly developed textbooks. However, they reported there was still a gap between the intended curriculum and the textbooks in the inclusion of specific content or in content arrangement. For example, real-life problem posing in the number and operation domain and decomposing natural numbers less than 20 were not included in elementary textbooks. They concluded that the intended curriculum should delimit the range and the scope of content with more explicit intent so that textbook writers would not have any confusion in making decisions on inclusion and arrangement of content.

Comparisons among Korean textbooks and those from other countries have actively attempted to reveal characteristics of Korean textbooks. For example, Son and Senk (2010) focused on how multiplication and division of fractions are developed in United States and Korean mathematics textbooks by comparing content, problems, and the number of lessons devoted to the topic. Both textbooks were found to provide opportunities to develop conceptual understanding and procedural fluency. A few differences in the two textbooks were found. In contrast to the U.S. textbook, in which conceptual understanding is introduced before algorithms for multiplication and division of fractions, the Korean textbook provided opportunities for both conceptual understanding and related procedural fluency side by side. Multistep computational problems are more common in the Korean textbook than in the U.S. textbook, and the response types are more varied in the Korean textbook (Son & Senk, 2010).

ENACTED MATHEMATICS CURRICULUM

As mentioned in the intended curriculum section, Korean mathematics teachers are expected to realize overall aims and objectives of the curriculum in their everyday teaching, an expectation that is related to the responsibilities and ethics of teachers. Research on everyday lessons by Korean mathematics teachers revealed, to some extent, how Korean mathematics teachers fulfill the intended curriculum. For example, Choe and Hwang (2004, 2005) investigated what Korean mathematics teachers set up as goals of their lessons, the range of learning content in their curriculum enactments, and how they set up these goals. The findings indicate that Korean mathematics teachers set up goals and taught content that were almost the same as what textbook writers suggested. Minor modification of phrases stating goals was sometimes done. In order to consider students' interest, content material was revised from time to time. Teachers tried reordering chapters or learning content rather than omitting or adding content in the adaptation of textbooks. Choe and Hwang (2005) argued that these prudent minor modifications of the mathematics curriculum, while looking for good strategies to orchestrate lessons, were what Korean mathematics teachers thought to be appropriate in their enactment of the intended curriculum.

Choe and Hwang (2004) found that first and second grade Korean mathematics teachers modified the textbooks by mainly considering students' achievement levels and the nature of the content to plan their lessons (see Table 9.3). Teachers in third grade differed in their responses. They not only considered students' achievement levels and the nature of the content but also modified the textbooks' teaching timeline or adjusted them to the local environment (see Table 9.3). Choe & Hwang (2004) emphasized the increase of mathematics teacher training on revised mathematics curriculum to develop their expertise on the mathematics curriculum.

Kim (2013) found the same tendency for secondary mathematics teachers to stick to textbooks when setting up goals and choosing content for teaching. Korean secondary mathematics teachers were found to use different teaching methods from those of the textbooks to address a variety of student achievement levels and their teaching environments. For example, mathematics teachers subdivided and added instructional objectives and contents of assessment to objects and content addressed in textbooks or guidebooks.

Research has repeatedly found that Korean mathematics teachers tend to use textbooks without changes in the goals and learning content but with some changes in teaching and learning methods. These tendencies show that Korean mathematics teachers fully respect and follow the written curriculum.

TABLE 9.3 Criteria for Reconstructing Mathematical Content in Curriculum Enactment

	Number of Responses (Percent)					
Characteristics of climate	26	(2.2)	25	(2.1)	57	(4.9)
Characteristics of region	56	(4.7)	81	(6.8)	160	(13.8)
Characteristics of contents	466	(38.9)	512	(43.2)	103	(8.9)
Level of students	597	(49.8)	397	(33.5)	77	(6.7)
Opinions of parents	4	(0.3)	16	(1.3)	70	(6.1)
School environment	27	(2.3)	54	(4.6)	140	(12.1)
Flexibility of school timetable	23	(1.9)	97	(8.2)	531	(45.9)
Etc.	0	(0.0)	4	(0.3)	19	(1.6)
Total	1,199	(100.0)	1,186	(100.0)	1,157	(100.0)

Source: Adapted from Choe and Hwang (2004)

LEARNED MATHEMATICS CURRICULUM

The learned mathematics curriculum can be described by the results of evaluation at the national level and the school level. The National Assessment of Educational Achievement (NAEA) is a national level evaluation annually administered by the Korea Institute for Curriculum and Evaluation (KICE). The results of the NAEA show to what degree the revised mathematics curriculum has been learned by students. Also, factors influencing student achievement have been investigated by analyzing the results of NAEA each year. Student achievement is classified into four levels: Advanced, Proficient, Basic, and Below Basic. In their analysis of tendencies revealed in the data from the NAEAs from 2010 to 2014, Lee, Cho, and Lee (2015) found that the percent of students in the Advanced and Basic levels decreased while the percent of those in the Proficient and Below Basic levels increased from 2013 to 2014. The scores of girls at the Proficient level increased while those of boys at the Proficient level decreased from 2013 to 2014. A greater number of boys than girls were at the Advanced and Below Basic levels but the reverse was true at the Proficient and Basic levels. Overall achievement improved but the achievement gap between schools in metropolises and small and medium-sized cities increased. The percent of students in rural areas who are at the Below Basic level has increased more than other areas; for example, there was a 1.66% increase in rural areas compared to 0.62% increase in metropolises from 2013 to 2014, although the government has made efforts to support both in their mathematics learning.

The College Scholastic Ability Test (CSAT), which is also conducted by KICE, indicates what and how Korean students learned the mathematics curriculum every year. Student achievement in the CSAT is classified into 9 levels. Kim and Shin (2010) report that students' aspiration toward learning, focused time on study, and teachers' perceptions of student learning during lessons are the main contributors to success in the CSAT. Private tutoring, learning strategies, learning motivation, and school environment are found to be partial contributors. In her analysis of the test items of CSATs from 2005 to 2011 with respect to the weight of each item, types of items, and the rate of correct answers, Nam (2011) suggested four changes that should be made in the CSAT. First, the difficulty level of CSAT should be lowered and test items should enhance learning motivation by being consistent with what the revised mathematics curriculum pursues. Second, short-answer items should be increased so that the CSAT can assess student learning rather than test-taking ability. Third, a sub-item system should be adopted to assess student learning in detail. Fourth, items that are solved by algorithm and repetitive practice should be reduced and new forms of items should be developed.

Korea is a country where parents have strong aspirations for their children's success in education, which usually means getting high scores on regular tests at school and on the CSAT. Because they are teaching children of this type of parent, Korean mathematics teachers and students are both under pressure to be well prepared for any kind of assessment. Teachers need to deal in their teaching with the types of problems that are often included on the tests and students need to practice solving problems within a short time. Thus, whenever curriculum revision is discussed, the assessment policy is also discussed in terms of what to change and how to change it. However, discussion of the assessment policy, especially regarding the university entrance examination, never goes smoothly. If there is any change, it has a huge impact on mathematics teaching and learning. Those who develop the mathematics curriculum are definitely aware of this sensitivity to assessment policy, so they include evaluation standards in the intended curriculum in relation to overall aims, objectives, and learning content. For example, there is a strong emphasis on so called process-oriented assessment, through which teachers can get information on the extent to which students have developed conceptual understanding, procedural fluencies, reasoning, communication skills, and so on.

Even though process-oriented assessment is emphasized in the intended curriculum, there is little evidence that it is enacted in practice (Chung, Lee, Yoo, Shin, & Kim, 2012; Park, 2010). According to Chung et al. (2012), teachers are aware that process-oriented assessment should be conducted but they are reluctant to practice it in their classrooms. The main reasons are difficulty in practicing due to lack of concrete methods and guidelines, concerns that come from the possibility of students' or parents' resistance to accept the results, and the fact that it is not related to high-stakes tests such as the CSAT. Therefore, it is a challenge to narrow the gap between the learned mathematics curriculum and the intended mathematics curriculum.

CHALLENGES AND CONCLUDING REMARKS

There are many challenges in Korea in the development and enactment of the mathematics curriculum, as is true in many other countries. Most of the challenges are deeply rooted in the Korean context of mathematics education, so they are not expected to change in a short time. We continuously tackle them with different strategies.

One challenge that has emerged is how to interpret the tendency for teachers not to alter the written curriculum. This could be seen as the written curriculum restricting teachers' autonomy to adapt learning content, even though they have students with different learning backgrounds. As reported in Choe and Hwang (2004, 2005) and Kim (2013), Korean mathematics teachers do not believe they need to or can alter the written curriculum. Instead, Korean mathematics teachers look for innovative teaching strategies different from those suggested in the teaching guides when considering students' variations in achievement and learning motivation. The tendency to stick to the written curriculum has not been interpreted as positive since it may reveal a lack of professional knowledge (i.e., Sherin & Drake, 2009). However, in the Korean context, sticking to the written curriculum can be interpreted as a contributor to narrowing the gap in student achievement among schools; whether they are located in metropolises or rural areas, students are given a similar amount of mathematical content by teachers. Thus, it is still an open question for further investigation into the background information of the Korean enacted curriculum, not only in how Korean mathematics teachers use curriculum, but also why they stick to following the written curriculum without adapting it.

The second challenge is bridging the gap between the general curriculum and the mathematics curriculum. We did not pay attention to this issue in this chapter but can briefly discuss it in relation to the development of the intended mathematics curriculum. The general curriculum is developed as a big umbrella for all subjects in Korea. In other words, when major reform ideas come from outside of the mathematics education community, they are interpreted and translated into mathematics education (Chang, 2014; Paik, 2004). It is natural to raise questions about why and how the mathematics curriculum realizes such reform ideas; major reform ideas often seem abstract or vague to mathematics curriculum developers. Thus, it has not been easy to interpret and translate reform ideas from the general curriculum into the mathematics curriculum. In addition, there is the moment when a reform idea comes to the surface and becomes a hot issue for discussion inside communities of mathematics educators. For instance, the contrast between Korean students' high performance and extremely negative disposition in mathematics as revealed in international comparative studies has stimulated rich discussion in and out of mathematics education, which has promoted some change in the intent of mathematics education. Hence, how to negotiate or accommodate the reform agenda that has emerged from both the general curriculum and mathematics curriculum sectors became very important. Currently, the KMOE handles this issue by giving a certain amount of independence to teams for curriculum development in each subject. The revised mathematics curriculum in 2015 was the result of this approach, although major reform ideas provided in the general curriculum are expected to be realized beforehand.

The major difficulties in bridging the gap between the general curriculum and the mathematics curriculum result from the fact that proper interpretations and meaningful translations of reform ideas in the general curriculum need considerable time for research, which has not been the case so far. For example, there need to be investigations and discussions on how to interpret the declared essential competences in the general curriculum for mathematics education. As we previously mentioned, there are six key competences that were translated from the general curriculum into the mathematics curriculum by mathematics educators: problem solving, communication, reasoning, creativity and convergence, attitude and practice, and data processing. These six competences were included in the previous curriculum rather than having been newly included. One might think that there is no change, but representation of certain ideas contained in the previous curriculum is repeatedly done by revision. Others might ignore or cannot see reform ideas because they look the same as ones in the previous curriculum. Therefore, we suggest that the Korean government give mathematics curriculum developers more freedom or autonomy so they can concentrate on how to include sophisticated reform ideas based on relevant research that originated in the context of Korean mathematics education.

The third challenge comes from the socio-cultural context of Korean mathematics education. The education fever of Korean parents has caused students to bear the burden of participating in private tutoring after school (Dawson, 2010). Pervasive private tutoring for extra mathematics learning is creating many problems, such as students' negative attitudes toward mathematics and financial crises in Korean families (Kim & Lee, 2010). There have been various voices dealing with this issue, but they have not come to an agreement or a solution. One of these voices has asked that the amount of content in the intended mathematics curriculum be decreased. Another has asked to change the assessment policy. Development of differentiated curriculum, extra support for slow learners, and the introduction of new teaching materials or methods have been suggested by various stakeholders in mathematics education. All the different voices seem to be having an impact, whether explicit or not or significant or not, on the revision and enactment of the mathematics curriculum. It is a challenge to harmonize these various voices so their impact on the Korean mathematics curriculum can be positive, or at least be handled.

In the previous sections, we briefly described how the Korean mathematics curriculum is developed, is enacted, and is learned or evaluated. We have observed that the written Korean mathematics curriculum has changed from being Confucian based to competence based in terms of overall pedagogical perspective. However, this change does not necessarily mean that Confucianism has been excluded from the curriculum. On the contrary, there are certain characteristics that may have originated from the Confucian approach, such as pursuing systemicity in the sense that learning content is vertically and horizontally systematized along the result from the research conducted to align the written curriculum with the enacted curriculum. This feature is related to the ethics and the responsibilities of textbook writers and teachers who carefully follow the intended curriculum so that the mission of public education can be fulfilled. Research to enhance the systemicity of curriculum has been conducted while facing challenges, such as continuous pressure to cut down the amount of content to be learned. For example, Yim, Lee, Lee, Park, & Jeong (2004) systematized learning goals and content through an in-depth investigation of the nature of school mathematics, deciding its optimal range and scope, drawing proper and effective ways of organizing school mathematics, and creating a sophisticated teaching and learning hierarchy. Through these efforts, the Korean mathematics curriculum is said to have a balance between a traditional and a reformed approach by integrating the main ideas shared among the international mathematics research community into Confucian-based perspectives. This kind of harmony will be our steady aspiration, though its results look different from generation to generation.

REFERENCES

Chang, H.-W., Kim, D.-W., & Lee, H.-C. (2013). Analysis on connection of curriculum and textbooks in elementary school mathematics: Focused on 1~2 grades. *School Mathematics, 15*(4), 759–783.

Chang, K. Y. (2014). Mathematics curriculum revising processes & directions from the standpoints of the contemporary two reports in the 1920s as the origin of math wars. *The Journal of Educational Research in Mathematics, 24*(4), 645–668.

Choe, S.-H., & Hwang, H.-J. (2004). A study on implementation of the seventh mathematics curriculum at the elementary school level. *School Mathematics, 6*(2), 213–233.

Choe, S.-H., & Hwang, H.-J. (2005). A study on the seventh national curriculum at the secondary school level. *School Mathematics, 7*(2), 193–219.

Chung, S.-K., Lee, K.-H., Yoo, Y.-J., Shin, B.-M., & Kim, G.-Y. (2012). *A study on process-focused assessment in school mathematics.* Seoul, Korea: Korea Foundation of the Advancement of Science & Creativity.

Dawson, W. (2010). Private tutoring and mass schooling in East Asia: Reflections of inequality in Japan, South Korea, and Cambodia. *Asia Pacific Education Review, 11*(1), 14–24.

Kang, W. (2013). An analysis on elementary mathematics curricula and textbooks of 2009 revised version in Korea: Four issues to be improved. *The Journal of Educational Research in Mathematics, 15*(3), 569–583.

Kheel, H.-S. (2002). A study on the learner-centered curriculum and instruction. *Journal of Learner-Centered Curriculum and Instruction, 7*(2), 1–26.

Kim, D.-H., Park, H.-S., Lee, J.-H., Kim, H.-J., Paik, S.-Y., Park, K., . . . & Lee, M.-H. (2009). *A study on the model of future-oriented mathematics curriculum focusing on creativity.* Seoul, Korea: Korea Foundation of the Advancement of Science & Creativity.

Kim, K.-S. (2007). A great leap forward to excellence in research at Seoul National University, 1994-2006. *Asia Pacific Education Review, 8*(1), 1–11.

Kim, M. H. (2013). Secondary mathematics teachers' use of mathematics textbooks and teachers' guide. *School Mathematics, 15*(3), 503–531.

Kim, S., & Lee, J. H. (2010). Private tutoring and demand for education in South Korea. *Economic Development and Cultural Change, 58*(2), 259–296.

Kim, Y.-B., & Shin, H.-S. (2010). Exploring student and school factors affecting student achievement on Korean CSAT (College Scholastic Ability Test): Focusing on reading and mathematics. *Journal of Educational Evaluation, 23*(3), 591–615.

Korean Ministry of Education [KMOE]. (2015a). *Mathematics curriculum. [Supplement 8]. Statute Notice of Ministry of Education* (No. 2015-74). Sejong, Korea: Korean Ministry of Education.

Korean Ministry of Education [KMOE]. (2015b). *Notes on publish and criteria for evaluating mathematics textbook for 2015 revised mathematics curriculum.* Sejong, Korea: Korean Ministry of Education.

Lee. I.-H., Cho, Y.-D., & Lee, K.-S. (2015). *An analysis of the result of National Assessment of Educational Achievement focusing on mathematics.* Seoul, Korea: Korea Institute for Curriculum and Evaluation.

Lee. K.-W., Min. Y.-S., Jeon, J.-C., Kim, M.-Y., & Kim, H.-J. (2008). *A study on developing key competencies in the primary/secondary school curriculum for the future of Koreans* (II). Seoul, Korea: Korea Institute for Curriculum and Evaluation.

Lew, H.-C. (1999). New goals and directions for mathematics education in Korea. In C. Hoyles, C. Morgan, & G. Woodhouse (Eds.), *Rethinking the mathematics curriculum* (pp. 218–227). London, England: Falmer Press.

Nam, J.-W. (2011). On the setting of mathematics test in the CSAT. *School Mathematics, 13*(1), 89–105.

Organization for Economic Cooperation and Development (OECD). (2002). *Education Policy Analysis 2002.* Paris, France: OECD.

Paik, S. Y. (2004). A study for the improvement of the way to reform mathematics curriculum. *The Journal of Educational Research in Mathematics, 14*(2), 157–170.

Park, K. (1997). A study on the three different versions of the Korean school mathematics. *The Journal of Educational Research in Mathematics, 7*(2), 91–101.

Park, K.-M. (2010). A comparative analysis of international mathematics curricula focusing on "grade band" and "mathematical process." *School Mathematics, 12*(4), 667–686.

Remillard, J. T., & Heck, D. J. (2014). Conceptualizing the curriculum enactment process in mathematics education. *ZDM: The International Journal on Mathematics Education, 46*, 705–718.

Sherin, M. G., & Drake, C. (2009). Curriculum strategy framework: Investigating patterns in teachers' use of a reform-based elementary mathematics curriculum. *Journal of Curriculum Studies, 41*(4), 467–500.

Shin, S.-K., Ko, J.-W., Kwon, J.-R., Park, S.-H., Lee, D.-H., Lee, B.-J., Choe, S.-H., & Cho, Y.-M. (2005). *A study on the improvement of the national school curriculum in mathematics.* Seoul, Korea: Korea Institute for Curriculum and Evaluation.

So, K. H. (2006). An investigation on new approaches to curriculum design for the knowledge-based society. *The Journal of Curriculum Studies, 24*(3), 39–59.

So, K., Kim, J., & Lee, S. (2012). The formation of the South Korean identity through national curriculum in the South Korean historical context: Conflicts and challenges. *International Journal of Educational Development, 32,* 797–804.

Son, J.-W., & Senk, S. L. (2010). How reform curricula in the USA and Korea present multiplication and division of fractions. *Educational Studies in Mathematics, 74*(2), 117–142.

Sung, B.-C. (2007). The conceptual framework and principles on educational publicness. *The Journal of Elementary Education, 20*(3), 229–249.

Yao, X. (2000). *An introduction to Confucianism.* New York, NY: Cambridge University Press.

Yim, J.-H., Lee, D.-H., Lee, Y.-R., Park, S.-K., & Jeong, Y. K. (2004). *Analysis and evaluation of the content relevance in the primary and secondary school mathematics.* Seoul, Korea: Korea Institute for Curriculum and Evaluation.

CHAPTER 10

TRANSCENDING BOUNDARIES

What Have We Learned?

Mary Ann Huntley
Cornell University

Christine Suurtamm
University of Ottawa

Denisse R. Thompson
University of South Florida

When we began conceptualizing this book on mathematics curriculum from an international perspective, we envisioned a volume in which we would learn more than simply, "countries A and B have a national mathematics curriculum, but countries C and D do not." At the outset, we offered a short list of issues for authors to consider when writing their chapter, and upon reading first drafts of the earliest chapters that were submitted, and considering issues raised by some of the authors, we suggested additional issues for them to address. However, we did not want *cookie cutter* chapters, with all authors addressing the same issues in the same order in lock-step fashion; rather, we wanted authors to feel free to tell the story of mathematics curriculum in their country in their own way. We are delighted that the authors did this, and openly shared what works, as well as the challenges and struggles regarding mathematics curricular issues in their countries. It should be noted that each chapter reflects the perspectives of the authors; in other words, a different set of mathematics educators in the country might have written a similar chapter, but with a different emphasis on certain aspects of the mathematics curriculum in their country.

We return now to the questions that were posed in Chapter 1, which were offered to help people focus their reading about mathematics curriculum in each of the eight countries represented in this book. We use information from the chapters to explore these questions. It is not our intent to provide an exhaustive set of responses. That is, we do not draw from all of the chapters in each set of responses; rather, for each question we provide illustrations from a subset of the countries represented in this book. In some cases, we provide comparative analyses in our responses to the questions; in others, we reflect on our own learning when reading the chapters.

How is Curriculum Defined in Each Country? What Does it Look Like?

The phrase *mathematics curriculum* is defined in different ways in various parts of the world. Moreover, there is no agreement on what constitutes *curriculum* or what words are used to denote *curriculum*. For instance, in Korea, there is a uniform curriculum that is administered at the national level. It includes content as well as pedagogical features such as teaching philosophy, learning principles, and evaluation. Although Canada has no national curriculum, the word *curriculum* refers to the sets of outcomes mandated by individual provincial ministries of education. In Brazil, the curriculum is based on prescriptive documents that define the content and teaching methods. In the United States, the word *curriculum* refers to a set of standards (e.g., *Common Core State Standards for Mathematics* or a specific state's standards), which contains content objectives as well as habits of mind, or ways of working, necessary for mathematical proficiency. It is often the case in the United States that people use the word *curriculum* to mean a textbook or a publisher's program. In the Netherlands, the *mathematics curriculum* for primary education is described in a *Core Goals* document, with an accompanying *Reference Framework* that provides more detail. In addition to content, these documents provide a view on learning, but do not focus on didactics; additional teaching-learning trajectories help teachers understand the processes needed to meet attainment targets. The curriculum in France is also presented in several official documents. One is called the *program*, which presents the content and associated skills that students should master, and other texts, called *accompanying resources*, offer detailed propositions with mathematical explanations. South Africa has a syllabus-type curriculum, which provides workbooks for students, as well as many resources for teachers, including clear specifications of the content, the order of teaching the topics in each grade, a brief summary of the principles underlying the approach, clarifying examples, descriptions of cognitive demand, sample lessons, and the duration of time to be allocated per topic. There is, however, some flexibility in that teachers may choose to sequence and pace the mathematical content differently from the recommendations. In Finland there is a national core curriculum, determined by the Finnish National Board of Education, that includes such things as general objectives and core content of each school subject, principles of student assessment, and teaching methods. So, in almost all countries the curriculum outlines the content to be addressed at various levels, but may not necessarily describe pedagogical perspectives related to that content. Also, there is variation across the countries in the extent to which curriculum documents are prescriptive.

What Does the Curriculum Development Process Look Like?

In the countries represented in this book, curriculum development varies widely in terms of who takes part in developing it, with what frequency, and with what motivation. In some countries, there appears to be a regular curriculum revision cycle. For example, in Finland there is an extensive curriculum revision approximately every 10 years. In other countries, there does not appear to be a regular cycle of revision. For example, in France, the mathematics curriculum has changed quite often since the early part of the 21st century, and seems to coincide with changes in government. As another example, South Africa's mathematics curriculum has had frequent revisions as a result of the end of apartheid and the results from different assessments impacting tertiary education.

In some countries, such as Canada, in which curriculum is not a national but a provincial responsibility, some provinces join together to develop a multi-jurisdictional curriculum. By joining forces, these provinces develop a curriculum and are able to work with publishers who otherwise might not be able to develop materials that align with curriculum in provinces with relatively small populations.

In many countries, curriculum development and revision involve a variety of stakeholders such as teachers, university mathematics professors, researchers in mathematics education, parents, and students. For instance, in developing the most recent core curriculum in Finland, teachers were to involve their students in planning and concretizing the curriculum. In Brazil, academics and university professors may help with curriculum revisions, but have little knowledge of classroom reality; thus, many teachers feel left out of the process and feel unprepared to implement curricular changes. In Western Canada, Ministry of Education teams develop programs of study, and then teachers with classroom experience work on developing curriculum materials.

What Role Does Research Play In Curriculum Development?

There is a growing body of research in mathematics education, much of which has the potential to shape and impact curriculum development, but such research is applied in different ways in different countries. For instance, in Brazil, two areas of research influenced the development of the National Curriculum Parameters to a large extent: studies on Ethnomathematics and problem solving. This curriculum document also incorporated research regarding history of mathematics, mathematical modeling, as well as information and communication technologies.

In the case of the Netherlands, research strongly impacts curriculum development. In 1997, the Ministry of Education commissioned the Freudenthal Institute to develop *teaching-learning trajectories* (*TAL*)—longitudinal teaching-learning trajectories with intermediate attainment targets—to be used in the curriculum. The main purpose of developing the TAL trajectories was to bring coherence in the school mathematics curriculum by providing a longitudinal overview of how children's mathematical understanding develops over the years, and outline how the different stages in this development are connected and build on each other. In practice, the intermediate attainment targets and teaching frameworks in the Netherlands curriculum form the essence of the intended teaching-learning processes.

In contrast to the heavy influence of research on curriculum development in Brazil and the Netherlands, in France the situation appears quite different: "In spite of the presence of educational researchers within the working groups, research results have not always been a central source for writing the official curriculum" (this volume, p. 48).

What Is the Underlying Framework of the Curriculum?

The assumptions and values that underlie a curriculum document set the tenor for what mathematics is valued, how students learn mathematics, and therefore how mathematics is taught in a country. For instance, the content of school mathematics being grounded in students' everyday experiences and learned with understanding are guiding forces for mathematics education in Brazil. In some countries, it appears that there is a struggle to maintain traditional perspectives, and at the same time, adopt ideas from research on how children learn mathematics. For example, basic mathematical knowledge and skills have historically been emphasized and are still valued in Korea, yet ideas from research on problem solving, and students engaging in reasoning, focusing on communication, and making connections across ideas, also currently influence the mathematics curriculum in that country.

In Chapter 7, the authors report that in South Africa the curriculum is founded on the principles of social transformation, active and critical learning, high knowledge and high skills, progression of content from simple to complex, and inclusivity. The authors of this chapter make a strong argument for including metacognitive ideology in the South African mathematics curriculum, as "in the absence of metacognition, the understanding of mathematics turns to memorizing facts and procedures, without a deep sense of awareness of the value of learning mathematics" (this volume, p. 182).

How Do Different Stakeholders View the Curriculum?

In some countries, instructional resources and assessments are the primary mechanisms by which expectations in official curriculum documents are communicated to teachers and parents. In such cases, as in the United States, these stakeholders often view the curriculum through textbooks and tests, rather than by directly reading official curriculum documents. By contrast, in Alberta (Canada), it has been observed that teachers frequently reference the program of studies when discussing mathematics lessons. It is not surprising that when changes are made to official curriculum documents, teachers and parents may become uneasy, especially when the changes are perceived to be rather large. For instance, when the Common Curriculum Framework was introduced in Alberta (Canada), teachers became unsettled, as the content and resources with which they were familiar were changing. Similarly, in the United States, in recent history there has been considerable controversy among teachers and parents concerning the content and teaching of mathematics, coinciding with changes in the school mathematics curriculum.

How are the Political, Cultural, Linguistic, Social, and Ideological Characteristics of the Country Reflected in the Curriculum?

The mathematics curriculum in a country reflects the views, norms, cultural, and historical context of the country within which it is situated. In some countries, the changes in curriculum have coincided with changes in politics. For example, in Chapter 3, Gueudet, Bueno-Ravel, Modeste, and Trouche (this volume) report that in France, "it is often said that each new government wants its own curriculum" (p. 42). They continue by saying, "the history of French mathematics curriculum is a tumultuous one that is very sensitive to scientific, social, and political tensions" (this volume, p. 43, Figure 3.1), which they chronicle since Napoleon's ordinance from the year 1802.

Changes to the mathematics curriculum in a country may reflect shifts in ideologies, as well as politics. Consider, for example, the case of Korea, where the curriculum historically was based on Confucianism, and has undergone significant changes as a result of the pro-democracy movement to establish a civilian government in the 1990s. Additionally, Lee, Park, and Ku (Chapter 9) report that a challenge that has arisen from the socio-cultural context of Korean mathematics education is parental pressure for students to participate in private after-school tutoring, which has led to problems, such as students' negative attitudes toward mathematics and financial difficulties for some families.

Another example of the mathematics curriculum being subject to influences outside of mathematics and education is South Africa, which experienced major curricular changes post apartheid. Additionally, there appear to be significant challenges in South Africa related to language, as there are 11 official languages of teaching and learning. In the United States, many aspects of the mathematics curriculum seem increasingly political, with a tug of war between federal, state, and local authorities.

The characteristics of a country may also have an influence on the mathematics curriculum. For example, in Finland, currently there are demands on curriculum from changes in Finnish society (e.g., immigration, an aging population, and new kinds of jobs). Another example is Brazil, where education authorities are challenged to establish standardized national guidelines in mathematics education because the country is geographically large and has enormous diversity.

What is the Role of Assessment on the Written, Intended, and Enacted Curriculum?

In some countries, changes in the written mathematics curriculum are heavily influenced by international assessments and their results. This appears to be the case in France, where it was found that the 2012 Programme for International Student Assessment (PISA) tests showed that students' achievements were highly correlated with their socio-economic backgrounds, which was an impetus for proposing a new curriculum in 2016. France's PISA test results have also been cited as one reason for the introduction in 2005 of the *common core* curriculum in France.

In other countries, such as Finland, results of international assessments seem to have little bearing with respect to the mathematics content of the school curriculum. In Finland, the mathematics curriculum is neither based on investigations into why Finnish students have historically had high scores on international assessments, nor into investigations as to why their scores are now trending downward. In Finland there is also the paradoxical situation of students performing, on international and national evaluations, quite poorly on algebra and competencies needed for the development of algebraic thinking, while at the same time, algebra prescriptions for Grades 1–6 have been relaxed.

In some countries, assessments influence teachers' decisions about the intended and enacted curriculum. For instance, in the United States, many teachers use high-stakes tests as the primary means of obtaining information about what is expected of students, which in turn provides information about the content they are expected to teach. Moreover, in the United States it is often the case that tasks that help prepare students for high-stakes assessments are routinely integrated into regular instruction. In this way, assessments sometimes drive teachers' content decisions in spite of what might be in the curriculum.

In still other countries, assessments at the end of one level of schooling, such as primary school, serve multiple purposes. In the Netherlands, the assessment at the end of Grade 6 provides insight into what level of secondary education a student will study, whether a student is at a basic or advanced level, and the overall quality of the school. So, this assessment has high value for students and school administrators.

What Does the Implementation of the Curriculum Look Like?

There is variation across countries, and also within countries, in terms of opportunities for teachers to learn about changes in their curriculum. For instance, in Chapter 4 it is reported that in Finland, although there are some materials provided by the Finnish National Board for Education for teachers to support curriculum reform, "there is no tradition of systematically supporting teachers' professional development or curriculum reforms on a large scale" (this volume, p. 89). Lack of support for curriculum implementation is also prevalent in Brazil, where it is reported in Chapter 8 that the official curriculum is often disconnected from the implemented curriculum, due to cultural, social, geographic, and economic differences across the (large) country.

Many factors that affect curriculum implementation have been discussed by the authors of chapters in this volume. For instance, in Canada, the applied mathematics stream that was introduced in 1998 ended up being discontinued just about the time it started to make headway, because teachers and counselors had not known how to advocate for it. As another example, many teachers in France are involved in the design of electronic resources that are coordinated by the Sésamath association, which is a group of practicing teachers. The success of these resources is having a major impact on implementation of the French curriculum. However, as stated in Chapter 3, "teachers are...expected to act as designers of their mathematics courses....At the same time, little inservice teacher education is offered" (this volume, p. 65). Similarly, in the United States, teachers are increasingly writing and sharing instructional resources on the Internet, thereby heavily influencing their selection of content and associated activities for use in classrooms. However, unlike the situation in France, in the United States this phenomenon is largely uncoordinated. In the United States this has resulted in concerns about curricular coherence, as "teachers find themselves navigating an uneven landscape, compiling a curriculum with insufficient support" (this volume, p. 160).

What is the Role of Textbooks and Other Resources in Enacting the Curriculum?

In some countries, textbooks have a strong influence on the enacted curriculum. For example, in France and the United States, textbooks and other instructional materials play a central role in mathematics classrooms. They influence the content that is taught, serve as resources for students, and provide teachers with support and guidance. But textbook adoption is handled differently in France than in the United States. In France, teachers are free to choose which textbook will be used by their students, whereas in the United States, decisions about which textbooks will be used in classrooms are often made by local school districts, or sometimes individual schools within these districts.

Likewise, in the Netherlands, textbooks have a major influence on classroom teaching. All of the major textbook series include "extensive teacher guidelines providing detailed information for each daily lesson, including directions for didactical approaches and differentiation" (this volume, p. 21). In addition, these teacher guides include content overviews for the grade as well as the expected learning goals.

Across the countries represented in this book, there is variability in the extent to which textbooks and resources align with the mathematics curriculum. Consider the case of South Korea, where mathematics textbooks are published only if the content aligns with the curriculum. To ensure this alignment, the Korean Ministry of Education has established criteria for evaluating mathematics textbooks. Textbooks can be published only if they satisfy these criteria, which are presented in the form of a detailed rubric.

What Is the Mathematical Focus of the Curriculum

The mathematical focus of a country's curriculum evolves over time, while at the same time it reflects the country's longstanding values and orientations toward mathematics. Consider France, which has an internationally recognized community of mathematicians. Here, rigorous proof has been in the past, and currently remains, an important aim of teaching. Although rigor is still a focus, the mathematics taught as part of the school curriculum now also includes more statistics, probability and analyses of variability, and computer science (e.g., algorithmics), all of which reflects the evolution of contemporary mathematicians' actual work.

In addition to focusing on the mathematical content, in some countries the curriculum also addresses mathematical processes, including students' dispositions. In Finland, for example, the school curriculum includes mathematical content and also outlines objectives for developing students' attitudes and self-efficacy. In all grade levels in Finland, mathematics instruction is designed to strengthen students' enthusiasm and interest for mathematics, and develop their self-confidence as a mathematics learner with the ability to take responsibility for their own learning.

In some countries, such as the Netherlands, particular attention is paid to curricular coherence, recognizing the importance of how all the elements of the curriculum work together, including the content, learning activities, aims and objectives, the teacher, materials and resources, and assessment. Using a *spider web model* as a visual illustration of coherence, curriculum researchers in the country recognize that curriculum is vulnerable, and changes in any one of these elements can influence how and what students learn.

SUMMARY AND FUTURE RESEARCH QUESTIONS

The chapters in this book provide a rich view of many aspects of the mathematics curriculum in eight countries: Brazil, Canada, Finland, France, the Netherlands, South Africa, South Korea, and the United States. These countries span five continents: South America, North America, Europe, Africa, and Asia. Given the geographical spread of the countries represented in this book, coupled with the fact that mathematics curriculum is heavily influenced by cultural traditions and political tendencies, the chapters provide diverse perspectives on a variety of issues surrounding mathematics curricula. This allows us to explore commonalities across the countries, and also to appreciate differences in approaches to, perspectives on, and practices concerning various issues regarding mathematics curriculum. Looking across geographic boundaries allows us to learn about what is working well in other countries, as well as to share in the struggles. There are often far more similarities in the challenges faced by educators in the classroom than might seem to be the case as reported in cross-national comparisons offered by policy makers.

We conclude this book with some questions that might warrant further investigation. Some questions have arisen from our own curiosity in reading the chapters, and others reflect issues/challenges that transcend national boundaries, as reported by authors of this volume.

1. How can teachers be better supported in understanding and implementing a new mathematics curriculum? Are there models for teacher professional development in some countries that are particularly effective in this regard that can be adopted or adapted to other contexts?
2. What aspects of curriculum contribute to student motivation and engagement? Are there methods that are particularly successful in some countries that can be adopted or adapted to other contexts?
3. In some countries there are frequent curriculum revisions, thereby limiting the time to test the effectiveness of a curriculum before another one is put in place. Are there mechanisms in some countries that are effective in using research to inform curriculum revision that can be adopted or adapted to other contexts? How is the implementation of revisions enhanced or hindered by assessment issues, particularly high-stakes assessments for graduation or entrance to tertiary-level education?
4. Teachers in many countries seem to be increasingly using the Internet to *design* their own curriculum, finding activities and problems that they perceive meet their curricular goals. Given this trend,

what are some ways that have been used successfully to support them to offer students a coherent mathematical experience?

5. To whom is the language used in writing a mathematics curriculum aimed? Are there mechanisms in some countries, which can be adopted or adapted to other contexts, for communicating mathematics curricular goals to stakeholders, including teachers, parents, and students, and to help add more coherence between the written, intended, and enacted curriculum?

We offer these questions with the hope that they stimulate communication among mathematics education leaders and curriculum researchers across countries, and also to provide some impetus for international collaboration on shared research projects. It is our goal that the chapters in this volume motivate readers to engage in discussions about mathematics curriculum development, implementation, and evaluation. We look forward to continuing the conversation.

ABOUT THE EDITORS

Mary Ann Huntley, PhD, is Senior Lecturer of Mathematics and Director of Mathematics Outreach and K–12 Education Activities in the Department of Mathematics at Cornell University in the United States. Her research involves examining the middle- and high-school mathematics curriculum from various perspectives, including the intended, enacted, and achieved curriculum, with a particular focus on the algebra strand of the curriculum. Awards for her scholarly work include a National Academy of Education Spencer Postdoctoral Fellowship and an American Association of Colleges of Teacher Education Outstanding Dissertation Award. She previously served as a program officer at the U.S. National Science Foundation, and has a background in applied mathematics. She is a co-editor of the series *Research in Mathematics Education*, published by Information Age.

Christine Suurtamm, EdD, is Professor of Mathematics Education at the Faculty of Education, University of Ottawa, Canada and teaches in the areas of mathematics education, assessment, and qualitative research. Her research focuses on the complexity of mathematics teachers' classroom practice, with particular interest in teachers' formative assessment practices as opportunities to attend to students' mathematical thinking. She is the Director of the Pi Lab, a research facility funded by the Canada Foundation for Innovation. She has been the principal investigator on several large-scale projects focusing on mathematics teaching and learning, was the Canadian representative on the National Council of Teachers of Mathematics (NCTM) Board of Directors, and was co-chair of Topic Study Groups on

Assessment at the past two International Congresses for Mathematics Education (ICME-12 & ICME-13). She has won several awards for research and teaching. She is a co-editor of the series *Research in Mathematics Education*, published by Information Age.

Denisse R. Thompson, PhD, is Professor Emeritus of Mathematics Education at the University of South Florida in the United States, having retired in 2015 after 24.5 years on the faculty. Her research interests include curriculum development and evaluation, with over thirty years of involvement with the University of Chicago School Mathematics Project. She is also interested in mathematical literacy, the use of children's literature in the teaching of mathematics, and in issues related to assessment in mathematics education; she served as co-chair of Topic Study Group 40 on Classroom Assessment at ICME-13. She is a co-editor of the series *Research in Mathematics Education*, published by Information Age.

ABOUT THE CONTRIBUTORS

Lætitia Bueno-Ravel is an associate professor in mathematics education at the teacher training institute of Bretagne, France (CREAD and ESPE de Bretagne, University of Brest, France, http://cread.espe-bretagne.fr/membres/Lbueno). Her research focuses on studying teachers' ICT integration in Mathematics Education, particularly at the primary school level. She studies the conditions and constraints for integrating new technologies in mathematics classrooms, considering new technologies as one of the resources, among many others, available for teachers. Using the documentational approach to didactics introduced by Ghislaine Gueudet and Luc Trouche, the issue of the articulation of new technologies and concrete resources is central in her work.

Regina Célia Grando received her undergraduate degree in mathematics, her master's degree in education and doctoral degree in mathematics education from "Universidade Estadual de Campinas" (UNICAMP/Campinas—São Paulo/Brazil). She is a professor and researcher in mathematics education at "Universidade Federal de Santa Catarina" (UFSC/Santa Catarina/Brazil) and is President of the "Sociedade Brasileira de Educação Matemática" (Brazilian Society of Mathematical Education). Her research interests are in the fields of teacher education, pedagogical practices, games in mathematics education, problem solving, and mathematics education in childhood.

Ghislaine Gueudet is a full professor in mathematics education at the teacher training institute of Bretagne, France (CREAD and ESPE de Bretagne, University of Brest, France, http://cread.espe-bretagne.fr/membres/ggueudet). Her research concerns several themes: university mathematics education, in particular the difficulties linked with the secondary-tertiary transition, and the design and use of educational resources (digital resources in particular). Concerning these resources, she has introduced in a joint work with Luc Trouche the documentational approach to didactics, analyzing teachers' interactions with resources and the consequences of these interactions in terms of professional development.

Kirsti Hemmi is a professor in mathematics and science education at Åbo Akademi University in Finland and visiting professor at Uppsala University in Sweden. Dr. Hemmi has led research projects comprising cultural aspects of mathematics education, curricula, textbooks, and teachers' interaction with materials, as well as projects aiming at school improvement at scale. She is currently leading a four-year research project focusing on teaching and learning of algebra at Uppsala University. She graduated as a class teacher from the University of Oulu in Finland in 1980 and later as a mathematics teacher from Stockholm University. She has a long experience of teaching mathematics at all school levels. Since 2001, she has also taught mathematics and mathematics education to prospective teachers. Dr. Hemmi earned her PhD in mathematics focusing on teaching and learning of proof at Stockholm University, Department of Mathematics, 2006.

Mary Ann Huntley, PhD, is Senior Lecturer of Mathematics and Director of Mathematics Outreach and K–12 Education Activities in the Department of Mathematics at Cornell University in the United States. Her research involves examining the middle- and high-school mathematics curriculum from various perspectives, including the intended, enacted, and achieved curriculum, with a particular focus on the algebra strand of the curriculum. Awards for her scholarly work include a National Academy of Education Spencer Postdoctoral Fellowship and an American Association of Colleges of Teacher Education Outstanding Dissertation Award. She previously served as a program officer at the U.S. National Science Foundation, and has a background in applied mathematics. She is a co-editor of the series *Research in Mathematics Education*, published by Information Age.

Divan Jagals, PhD, completed his thesis at North-West University, Potchefstroom Campus, South Africa, where he is a lecturer in Curriculum Studies, Philosophy and Research Methodology. His research interests range from theory to philosophy of education with a focus on the application of metacognition on the philosophy of mathematics education. Specifically, he explores the philosophical theories of ontology, epistemology, methodology as well as teleology, and their relationship with the teaching and learning of mathematics. He is a co-investigator in two nationally funded research projects on the affordances of indigenous knowledge for self-directed learning and has presented and published papers nationally and internationally.

Heidi Krzywacki, PhD, works as University Lecturer and Researcher in mathematics education at the Department of Teacher Education, University of Helsinki, Finland. She worked as a teacher educator in Mälardalen University in Sweden from 2013–2015. She is originally a qualified primary school teacher as well as a mathematics teacher. Her doctoral thesis focused on the process of becoming a teacher during preservice mathematics teacher education, which was conceptualized through the professional identity of teachers. Recently, her research interest has been on the relation between curriculum materials and teachers in mathematics classrooms as well as on school assessment. She was a member of the working group on the reform of the national core curriculum for pre-primary and basic education mathematics in 2014 (Finnish National Board of Education).

Na-Young Ku is a mathematics teacher in South Korea. She has taught high school mathematics for nearly 10 years. She has a master's degree in mathematics education and is a doctoral student in mathematics education at Seoul National University. Her research interests are the enacted curriculum, discourse analysis in mathematics classes, probability and statistics education, and teacher professional development. Her most recent research agenda involves analysis of factors that influence teachers' curriculum use and how teachers use and enact mathematics curriculum materials in mathematics classrooms. She has participated in several research projects, such as *Research on Improvement of Mathematics Curriculum for Activation of Statistics Education Project* and *Supporting Professional Learning Community of Mathematics Teachers Project*.

Kyeong-Hwa Lee is a full professor at Seoul National University in South Korea. Her research and teaching interests embrace curriculum development and curriculum evaluation, education and assessment of mathematical creativity, textbook task analysis and modification, discourse analysis in mathematics classes, probability and statistics education, analogical reasoning, gender issues, preservice teacher education, and teacher professional development. She has been in charge of several research projects, such as the *Women into Science and Engineering Project*, *Advance Planning for Development of Mathematics Textbook Project*, and *Supporting Professional Learning Community of Mathematics Teachers Project*. She has published more than 80 journal articles, books, book chapters, reviews, and commentaries in mathematics education.

Celi Espasandin Lopes is a professor of mathematics education in the graduate program in mathematics and science teaching at "Universidade Cruzeiro do Sul" (UNICSUL/São Paulo/Brazil). She graduated with degrees in mathematics and pedagogy and obtained both her master's and doctoral degrees in education from "Universidade Estadual de Campinas" (UNICAMP/Campinas—São Paulo/ Brazil). She was in a visiting/post-doctoral position in mathematics education at The University of Georgia (UGA) and Miami University in the United States. Dr. Lopes is the coordinator of the Center for Research in Mathematics Education and Statistics (CEPEME) and is the lead faculty in the Study Group for Research in Statistics and Mathematics Education (GEPEEM). Her research interests are in mathematics and statistics education.

Simon Modeste is an associate professor in didactics of mathematics at the *Institut Montpelliérain Alexander Grothendieck* (IMAG, University of Montpellier, France, http://www.math.univ-montp2.fr/~modeste/). During his PhD, he studied the epistemological and didactical issues involved in the teaching and learning of algorithmics, and the particular case of the introduction of algorithmics in the high school mathematics curriculum in France. Now, his research is dedicated, more generally, to the epistemological and didactical questions that arise from the interactions between the disciplines of informatics and mathematics. In his university, he takes part in the preservice and inservice training of mathematics teachers.

JinHyeong Park is an assistant professor at Myongji University, South Korea. His research and teaching interests include mathematical modelling, curriculum development and revision, textbook development and analysis, mathematical creativity and giftedness, examples and exemplification, abductive reasoning, generalization, semiotic and discursive analysis in mathematics classes, and teacher professional development in mathematics education. He has also participated in several research projects, such as the *Supporting Professional Learning Community of Mathematics Teachers Project*, *Lesson Planning for Mathematics Project-based Learning*, *Facilitation of Creativity in Mathematics Evaluation for University Admission Project*, and *Process-Focused Assessment in School Mathematics Project*. He has published many journal articles and book chapters in mathematics education.

Anna-Maija Partanen, PhD, works as a senior university lecturer in the teacher training section of the Faculty of Education at the University of Lapland, Finland. She also leads the LUMA-Centre Lapland, that is a part of a network of 13 STEM-Centers located in different Finnish universities. She has a long experience as an upper secondary mathematics teacher. In 2011 she defended her thesis, which focused on social and sociomathematical norms, describing a tension when an investigative small-group method was used in one of her advanced-mathematics classes. She participated in the *Nordic-Californian VIDEOMAT-project* in 2011–2013, and currently leads a development and research project in *Early Algebra*, the aim of which is to design and test teaching materials for Finnish elementary schools.

Luke Reinke, PhD, is an assistant professor of mathematics education in the Cato College of Education at the University of North Carolina at Charlotte in the United States. He teaches courses in mathematics teaching methods and instructional technology. His research focuses on the use of contextualized problems in mathematics instruction and teachers' use of mathematics curriculum materials. His doctoral dissertation describes how the role of contextualized problems shifted as a teacher and her students enacted a modeling-based algebra unit designed to align with NCTM's *Principles and Standards for School Mathematics*. He is also interested in curriculum development and has significant experience designing digital instructional resources for students and teachers.

Janine Remillard, PhD, is an associate professor of mathematics education at the University of Pennsylvania's Graduate School of Education (Penn-GSE) in the United States. Her research interests include teachers' interactions with mathematics curriculum materials, mathematics teacher learning in urban classrooms, and locally relevant mathematics instruction. She is one of the primary faculty in Penn-GSE's urban teacher education program and is co-editor of the volume, *Mathematics Teachers at Work: Connecting Curriculum Materials and Classroom Instruction*. Remillard has undertaken research and development projects on teacher learning, curriculum use, and formative assessment. She is active in the mathematics education community in the United States and internationally, having chaired the U.S. National Commission on Mathematics Instruction, a commission of the National Academy of Science, and served two terms on the board of the AERA SIG-Research in Mathematics Education. She is also involved in international, comparative research on mathematics curriculum, currently collaborating with researchers from Belgium, Sweden, and Finland.

Elaine Simmt is Professor of Secondary Education (mathematics) at the University of Alberta, Canada. She began her career teaching secondary school mathematics and physical sciences. She teaches and researches in the fields of mathematics education and mathematics teacher education.

Elaine has been investigating teaching and learning in classrooms and professional development contexts using the theoretical frames of enactivism and complexity. Her work has included studying the implications of high activity mathematics lessons, the mathematics needed for teaching, and most recently she is working with colleagues trying to identify methods and techniques for studying collective learning. International teacher professional development work in South Africa, Tanzania, and Oman have triggered her interest in the history of mathematics curriculum and teacher education in Canada.

Christine Suurtamm, EdD, is Professor of Mathematics Education at the Faculty of Education, University of Ottawa, Canada and teaches in the areas of mathematics education, assessment, and qualitative research. Her research focuses on the complexity of mathematics teachers' classroom practice, with particular interest in teachers' formative assessment practices as opportunities to attend to students' mathematical thinking. She is the Director of the Pi Lab, a research facility funded by the Canada Foundation for Innovation. She has been the principal investigator on several large-scale projects focusing on mathematics teaching and learning, was the Canadian representative on the National Council of Teachers of Mathematics (NCTM) Board of Directors, and was co-chair of Topic Study Groups on Assessment at the past two International Congresses for Mathematics Education (ICME-12 & ICME-13). She has won several awards for research and teaching. She is a co-editor of the series *Research in Mathematics Education*, published by Information Age.

Denisse R. Thompson, PhD, is Professor Emeritus of Mathematics Education at the University of South Florida in the United States, having retired in 2015 after 24.5 years on the faculty. Her research interests include curriculum development and evaluation, with over thirty years of involvement with the University of Chicago School Mathematics Project. She is also interested in mathematical literacy, the use of children's literature in the teaching of mathematics, and in issues related to assessment in mathematics education; she served as co-chair of Topic Study Group 40 on Classroom Assessment at ICME-13. She is a co-editor of the series *Research in Mathematics Education*, published by Information Age.

Luc Trouche is a full professor in mathematics education at the French Institute of Education (Ecole Normale Supérieure de Lyon, France; https://ens-lyon.academia.edu/LucTrouche). His research has been dedicated to studying Information and Communications Technology (ICT) integration in Mathematics Education. In particular, he has studied the interplay between instrumentation processes and conceptualization processes. This work led him to study the teacher's role, introducing the notion of orchestration for modeling the management of available artifacts (for teaching a particular mathematical topic) in the classroom. This has led him to introduce, in a joint work with Ghislaine Gueudet, the documentational approach to didactics, analyzing teachers' interactions with resources and the consequences of these interactions in terms of professional development.

Marja van den Heuvel-Panhuizen is Full Professor of Mathematics Education at Utrecht University and is affiliated both with the Freudenthal Institute of the Science Faculty and with the Freudenthal Group of the Department of Education and Pedagogy of the Faculty of Social and Behavioural Sciences in the Netherlands. Since 2016, she is emeritus professor. She started her work as a researcher at the Freudenthal Institute in 1987 and was a visiting professor at Dortmund University and at IQB of Humboldt University, Berlin. Since May 2017, she also has a part-time full professorship at Nord University, Norway. She was and is involved in numerous national and international projects on primary school mathematics, special education, and early childhood. Among other things, she was the project leader of a national government-funded project on developing teaching-learning trajectories for primary school mathematics. Currently she is involved in studies on assessment, textbook analyses, picture book use in mathematics education, and mathematics-related higher-order thinking in primary school students.

Marthie van der Walt is an associate professor in mathematics education, in the Faculty of Education Sciences at the North-West University, South Africa. Her research forms part of the focus area for self-directed learning. Her research interests include metacognition (including reflection) in mathematics teacher education (Grades 4 to 9), lesson study and professional development (pedagogical content knowledge [PCK] of teachers), study orientation in mathematics (triMATHS questionnaires for use in Grades 4 to 7), contextualised mathematics, and case-based teaching. Her publications appear in national and international journals. Her current participation as co-investigator in funded research projects includes *The affordances of indigenous knowledge for self-directed learning* and *Teachers without borders*. Presently, she is involved in training preservice mathematics teachers and supervising post-graduate students in mathematics education, focusing on metacognition.

Marc van Zanten, Drs., is a PhD student at the Freudenthal Institute of the Science Faculty and the Freudenthal Group of the Faculty of Social and Behavioral Sciences of Utrecht University, the Netherlands. He is a former primary school teacher and teacher educator, specializing in primary school mathematics education. He was chair of the committee that developed the

Dutch Knowledge Base on Mathematics for primary school teachers. Currently, he works as a researcher and developer of mathematics education at the Netherlands Institute for Curriculum Development (SLO). Since 2007, he has served as Chair of the Dutch annual Panama-conference on mathematics education for researchers, educators, and school advisors. In his PhD research, he investigates textbooks for primary school mathematics education, focusing on the learning opportunities that textbooks offer.

Made in the USA
Middletown, DE
02 March 2018